... is Series

Titles in This Series

Titles in This Series

Titles in This Series

Algebraic Geometry: Sundance 1988

CONTEMPORARY
MATHEMATICS

116

Algebraic Geometry: Sundance 1988

Proceedings of a Conference on Algebraic Geometry
held July 18–23, 1988
with support from Brigham Young University
and the National Science Foundation

Brian Harbourne
Robert Speiser
Editors

American Mathematical Society
Providence, Rhode Island

The Conference on Algebraic Geometry was held at Sundance, Utah on July 18–23, 1988 with support from Brigham Young University and the National Science Foundation, Grant DMS-8719233.

1980 *Mathematics Subject Classification* (1985 *Revision*). Primary 14C05, 14D20, 14H10, 14C20, 14C22, 14J26, 14E30, 14B05, 14N05, 14C17, 14B07, 14N10, 14J25, 14E25, 14E17.

Library of Congress Cataloging-in-Publication Data

The Conference on Algebraic Geometry (1988: Sundance, Utah)
Algebraic geometry: Sundance 1988: proceedings of a Conference on Algebraic Geometry held July 18–23, 1988/Brian Harbourne and Robert Speiser, editors.
    p. cm.—(Contemporary mathematics; 116)
    ISBN 0-8218-5124-1 (alk. paper)
    1. Geometry, Algebraic—Congresses. I. Harbourne, Brian, 1955– . II. Speiser, R. (Robert) III. Title. IV. Series: Contemporary mathematics (American Mathematical Society); v. 116.
QA564.C657 1988
516.3'5—dc20

90-29884
CIP

Contents

Preface

This volume represents the proceedings of the 1988 Sundance conference on algebraic curves and related varieties, July 18–23, 1988. Some papers here represent actual conference lectures, or report research carried out at Sundance, while others describe related work done elsewhere.

The beauty of the mountains, the fine weather, the particularly comfortable working environment, and the participants' energy combined to make this conference especially memorable. It is a pleasure to thank everyone who took part and express gratitude as well to Brigham Young University and the National Science Foundation (grant DMS-8719233) for strong support.

Brian Harbourne
Robert Speiser

Contemporary Mathematics
Volume **116**, 1991

Hurwitz Surfaces with Nontrivial Divisors

STEVEN DIAZ AND RON DONAGI

Introduction

Hurwitz schemes parametrize families of branched covers of $\mathbf{P}^1$. A Hurwitz surface is a Hurwitz scheme of dimension 2. At the 1988 Sundance conference it was conjectured (not by any of the authors) that the group of divisors modulo algebraic equivalence on a Hurwitz scheme is torsion. Here we provide a counter-example to this conjecture by constructing a Hurwitz surface that has a divisor no positive multiple of which is algebraically or even homologically equivalent to zero. We also make some comments about possible new conjectures and talk about the problem of determining which varieties of arbitrary dimension are Hurwitz schemes.

§1 gives a precise statement of the conjecture and some motivation for believing it. In §2 we construct a family of counter-examples to the conjecture. §3 states some possible new questions one could ask. Finally in §4 we discuss the problem of determining which varieties are Hurwitz schemes.

1

The Hurwitz scheme $H_{k,b}$ is the coarse moduli space for isomorphism classes of irreducible k-sheeted branched covers of $\mathbf{P}^1$ whose branch locus consists of exactly b distinct ordered points of $\mathbf{P}^1$, cf., [F]. Three of these branch points are fixed at 0, 1 and ∞; the remaining $b-3$ are allowed to vary. One therefore has a morphism:

(1)
$$\pi_{k,b} : H_{k,b} \to (\mathbf{P}^1)^{b-3} \backslash S$$
$$(f : C \to, \mathbf{P}^1) \to (\text{branch } f \backslash \{0, 1, \infty\}),$$

where S consists of all diagonals and all points involving 0, 1 or ∞. Standard references such as [F] show that $\pi_{k,b}$ is finite and étale.

1980 *Mathematics Subject Classification* (1985 *Revision*). Primary 14C05, 14D20, 14H10.
This paper is in final form and no version of it will be submitted for publication elsewhere.
The second author was partially supported by National Science Foundation grant DMS87-12298.

Let us fix a connected component H of some $H_{k,b}$. All curves appearing as covers corresponding to points of H have the same genus g, where g is determined using the Riemann-Hurwitz formula. Letting $\mathcal{M}_g$ be the (coarse) moduli space of isomorphism classes of curves of genus g we get a morphism:

$$\sigma : H \to \mathcal{M}_g$$
(2)
$$(f : C \to \mathbf{P}^1) \to (\text{moduli point of } C),$$

Continue with a fixed component H. We want to know the group of divisors on H modulo either linear, algebraic, or homological equivalence. It was once conjectured that these groups were all torsion, for any H. There are a number of reasons to suspect that this is true.

First consider the map $\pi_{k,b}$ of (1). The group of divisors modulo linear equivalence on $(\mathbf{P}^1)^{b-3}\backslash S$ is zero. Thus, one does not find any nontrivial divisors on H by pulling back divisors from $(\mathbf{P}^1)^{b-3}\backslash S$. Next consider the map of (2). The only nontrivial divisors on $\mathcal{M}_g$ $(g \geq 3)$ are multiples of the Hodge divisor λ, see [AC, H]. Using Porteous-type calculations similar to those found in [DH] it is not hard to show that (in many cases—perhaps always) $\sigma^*\lambda$ is torsion. Again no nontorsion divisors are found. Finally, again using Porteous-type calculations as in [DH], it is not hard to show that any geometrically defined divisor we could think of on H is torsion.

Despite all this evidence the conjecture is false. For divisors modulo linear equivalence many counter-examples are provided by the main result of [DDH]. There it is shown that H could be isomorphic to a Zariski open subset of an irrational curve. For divisors modulo algebraic or homological equivalence we provide counter-examples in the following section.

2

The construction in this section is based on constructions from [DDH]. We start by reviewing material from [DDH] that we will need. We omit proofs which can be found in [DDH].

Construction (X). Let X be a complete nonsingular curve, $S \subset X$ a finite set, and $n > 0$. We have a surjection

$$\alpha : \pi_1(X\backslash S) \twoheadrightarrow H_1(X\backslash S, \mathbf{Z}) \twoheadrightarrow H_1(X\backslash S, \mathbf{Z}/n\mathbf{Z}),$$

(where the fundamental group is with respect to a base point which we do not specify.) Since $\ker(\alpha)$ is a normal subgroup of finite index in $\pi_1(X\backslash S)$, it determines a finite Galois cover of $X\backslash S$ and a corresponding branched cover $g : X' \to X$. We note that g is:

  (a) unramified, if S consists of a single point, and
  (b) ramified with ramification index n at each point of $g^{-1}(S)$, if $\text{Card}(S) > 1$.

Fix a branched cover $\pi : C \to \mathbf{P}^1$ with branch locus contained in a finite subset B of $\mathbf{P}^1$. For each point $p \in C \backslash \pi^{-1}(B)$ we will now apply Construction (X) to obtain a cover $f_p : C_p \to \mathbf{P}^1$ with branch locus equal to $B \cup \{\pi(p)\}$.

Construction (C). Fix the following data: a point z of $\pi^{-1}(B)$, a cover $h : \tilde{C} \to C$ unbranched outside $\pi^{-1}(B) \backslash z$, a subset S' of $\tilde{C}$ consisting of one point over each point of $B \backslash \pi(z)$ over which $\pi \circ h$ fails to branch, and one point y of $h^{-1}(z)$. In Construction (X) take $X := \tilde{C}$ and $S := h^{-1}(p) \cup S' \cup \{y\}$. Rename the resulting cover $g : X' \to X$ as $g_p : C_p \to \tilde{C}$, and let $f_p = \pi \circ h \circ g_p$. Note that g_p is ramified with ramification index n at each point of $g^{-1}(S)$, since $\mathrm{card}(S) > 1$. This is similar to Parshin's construction, [**P**, p. 1163].

3. LEMMA. *Let $\tilde{f} : \tilde{X} \to X$ be a branched cover, where X and $\tilde{X}$ are nonsingular projective curves. Let T be a finite set of points of $\tilde{X}$. Suppose that X_0 and X_1 are intermediate covers, i.e., $\tilde{f} : \tilde{X} \xrightarrow{g_i} X_i \xrightarrow{f_i} X$ $(i = 0, 1)$ and let $p_i \in X_i$. Assume that $g_i^{-1}(p_i) = T$ $(i = 0, 1)$ and that f_i is unramified at p_i. Then X_0 and X_1 are isomorphic as intermediate covers of $\tilde{f} : \tilde{X} \to X$.*

4. LEMMA. *Let $\pi : C \to \mathbf{P}^1$ have branch locus contained in $B \subset \mathbf{P}^1$ where $\mathrm{Card}(B) \geq 3$. Then there exists a branched cover $h : \tilde{C} \to C$ with the following properties:*

(1) *The branch locus of h is contained in $\pi^{-1}(B)$.*
(2) *There is at least one point of $\pi^{-1}(B)$ over which $\tilde{C} \to C$ is unramified.*
(3) *No automorphism of $C \to \mathbf{P}^1$, other than the identity, lifts to an automorphism of $\tilde{C} \to \mathbf{P}^1$.*

This concludes the review of material from [**DDH**].

Fix a branched cover $\pi : C \to \mathbf{P}^1$ with branch locus contained in $\{0, 1, \infty\}$. We will show that there is a component H of some Hurwitz scheme $H_{k,5}$ such that H is isomorphic to a Zariski open subset of the product $C \times C$. For notational purposes set $(\mathbf{P}^1 \times \mathbf{P}^1)^0 := \mathbf{P}^1 \times \mathbf{P}^1 \backslash$ (the diagonal and all points involving $0, 1, \infty$) and $(C \times C)^0 := (\pi \times \pi)^{-1}(\mathbf{P}^1 \times \mathbf{P}^1)^0$.

Construction $(C \times C)$. Using (4) fix a branched cover $h : \tilde{C} \to C$ satisfying the requirements of Construction (C) with $B = \{0, 1, \infty\}$ and such that no automorphism of C over $\mathbf{P}^1$ lifts to an automorphism of $\tilde{C}$ over $\mathbf{P}^1$. Fix two distinct prime numbers l, m. For each $(p, q) \in (C \times C)^0$ we construct a branched cover $f_{p,q} : C_{p,q} \to \mathbf{P}^1$ branched exactly at $\{0, 1, \infty, \pi(p), \pi(q)\}$ as follows. Let $f_p : C_p \to \mathbf{P}^1$ $(f_q : C_q \to \mathbf{P}^1)$ be as in Construction (C) with $n = l$ $(n = m)$ in Construction (X). Set $C_{p,q}$ equal to the fiber product

$C_p \times_{\tilde{C}} C_q$ and set $f_{p,q}$ to be the composition of the natural projection of $C_{p,q}$ to C_p (or C_q) with f_p (or f_q). The fact that $C_{p,q}$ is a nonsingular connected curve follows from the fact that l and m are relatively prime.

This construction may be carried out with the same *fixed* data for each point (p, q) of $(C \times C)^0$. There is also an ordering of the branch points on $\mathbf{P}^1$: $0, 1, \infty, \pi(p), \pi(q)$. We thereby get a map $\omega : (C \times C)^0 \to H_{k,5}$ (some k) taking (p, q) to the point of $H_{k,5}$ corresponding to the cover $f_{p,q} : C_{p,q} \to \mathbf{P}^1$. We therefore have a commutative diagram:

$$(5) \qquad
\begin{array}{ccc}
(C \times C)^0 & \xrightarrow{\;\;\omega\;\;} & H_{k,5} \\
 & & \\
\pi \times \pi \searrow & & \swarrow \pi_{k,5} \\
 & (\mathbf{P}^1 \times \mathbf{P}^1)^0 &
\end{array}$$

From the fact that $(C \times C)^0$ is a proper over $(\mathbf{P}^1 \times \mathbf{P}^1)^0$ it follows that ω is surjective onto some component H of $H_{k,5}$. Since both $(C \times C)^0$ and H are nonsingular surfaces, to show that ω is an isomorphism onto H it is sufficient to show that ω is one-one.

If ω is not one-one we have an isomorphism $\alpha : C_{p,q} \to C_{p',q'}$ over $\mathbf{P}^1$ for some $(p, q) \neq (p', q')$. We may as well assume $p \neq p'$. Let π_p and $\pi_{p'}$ be the natural projections of $C_{p,q}$ to C_p and $C_{p',q'}$ to $C_{p'}$. We claim that in this case one may fill in the dotted arrows to make the following diagram commutative

$$(6) \qquad
\begin{array}{ccc}
C_{p,q} & \xrightarrow{\quad\alpha\quad} & C_{p',q'} \\
g_p \circ \pi_p \downarrow & & \downarrow g_{p'} \circ \pi_{p'} \\
\tilde{C} & \dashrightarrow{\;\beta\;} & \tilde{C} \\
h \downarrow & & \downarrow h \\
C & \dashrightarrow{\;\gamma\;} & C \\
 & \pi \searrow \quad \swarrow \pi & \\
 & \mathbf{P}^1 &
\end{array}$$

in such a way that $\gamma(p) = p'$ and $\gamma(q) = q'$. To construct γ, we apply (3) to:

$$
\begin{array}{ccc}
 & C_{p,q} & \\
g_0 := h \circ g_p \circ \pi_p \;\swarrow & & \searrow\; g_1 := h \circ g_{p;} \circ \pi_{p'} \circ \alpha \\
C & \dashrightarrow{\;\gamma\;} & C \\
\pi \searrow & & \swarrow \pi \\
 & \mathbf{P}^1 &
\end{array}$$

The lemma applies since, $g_0^{-1}(p) = g_1^{-1}(p') =$ the ramification points of $C_{p,q} \to \mathbf{P}^1$ that have ramification index l (l is a prime fixed in Construction $(\tilde{C} \times C)$) and do not lie over $\{0, 1, \infty\}$. The resulting γ takes Branch (g_0) to Branch (g_1), hence p to p' and q to q'.

To construct β we similarly apply (3) to:

$$
\begin{array}{ccc}
& C_{p,q} & \\
g_p \circ \pi_p \nearrow & & \searrow g_{p'} \circ \pi_{p'} \circ \alpha \\
\widetilde{C} & \xdashrightarrow{\;\;\beta\;\;} & \widetilde{C} \\
\gamma \circ h \searrow & & \nearrow h \\
& C &
\end{array}
$$

this time using the equality $(g_p \circ \pi_p)^{-1}(y) = (g_{p'} \circ \pi_{p'} \circ \alpha)^{-1}(y) = $ (the ramification locus of $C_{p,q} \to C) \cap$ (fiber over z) and the assumption (in Construction (C)) that h is unramified over z.

The existence of diagram (6) contradicts the assumption that no automorphism of C over $\mathbf{P}^1$ lifts to an automorphism of $\widetilde{C}$ over $\mathbf{P}^1$. Thus we conclude that ω is one-one and hence an isomorphism onto H.

To construct the desired counter-example we apply Construction $(C \times C)$ to an appropriate choice of $\pi : C \to \mathbf{P}^1$. Consider the modular curves $X_0(N)$. A general point of $X_0(N)$ corresponds to a pair (E, G) where E is an elliptic curve (with a marked origin) and G is a cyclic subgroup of E of order N. The j-invariant gives a map from $X_0(N)$ to $\mathbf{P}^1$ all of whose ramification points lie over $\{0, 1, \infty\}$.

(7) PROPOSITION. *In Construction* $(C \times C)$ *take* $\pi : C \to \mathbf{P}^1$ *to be* $j : X_0(N) \to \mathbf{P}^1$. *Let* $H := (X_0(N) \times X_0(N))^0$ *be the resulting component of a Hurwitz scheme, that is, the image of* ω *in* (5). *Then for appropriate choices of* N, H *will contain a divisor no positive multiple of which is either algebraically or homologically equivalent to zero.*

PROOF. Any divisor algebraically equivalent to zero is certainly homologically equivalent to zero, so it is enough to prove this for homological equivalence. The relevant group to study is thus the Hodge group

$$H^{1,1}(X_0(N) \times X_0(N), \mathbf{C}).$$

The proposition is immediate once we have proven the following two claims.

CLAIM 1. For appropriate choices of N, the dimension of the $\mathbf{C}$ vector space $H^{1,1}(X_0(N) \times X_0(N), \mathbf{C})$ is at least 4.

CLAIM 2. For any N, the dimension of the subspace of

$$H^{1,1}(X_0(N) \times X_0(N), \mathbf{C})$$

spanned by divisors supported on $X_0(N) \times X_0(N) \backslash (X_0(N) \times X_0(N))^0$ is at most 3.

PROOF OF CLAIM 1. We use standard knowledge about modular curves, cf., [BK]. For N where $X_0(N)$ has positive genus we have at least 3 independent divisor classes on $X_0(N) \times X_0(N) : p \times X_0(N)$, $X_0(N) \times q$, and the diagonal. Now consider the graph of a Hecke correspondence T on $X_0(N) \times X_0(N)$. As a correspondence T induces an endomorphism of the Jacobian of $X_0(N)$ and thus a linear operator on the cotangent space to the origin of this Jacobian. This linear operator is the action of the Hecke operator on weight two cusp forms. If the graph of the Hecke correspondence were homologous to a linear combination of the other three mentioned divisors then this induced linear operator would be a multiple of the identity. For many N this is not the case: Table 5 in [BK] shows that the involution W_2 has 2 distinct rational eigenspaces in $X_0(22)$, and that the Hecke operators T_p have 2 distinct rational eigenspaces on $X_0(26)$.

PROOF OF CLAIM 2. Consider the map

$$j \times j : X_0(N) \times X_0(N) \to \mathbf{P}^1 \times \mathbf{P}^1.$$

Clearly the inverse image of anything of the form, point $\times \mathbf{P}^1$, or $\mathbf{P}^1 \times$ point, is homologous to a multiple of, point $\times X_0(N)$, or $X_0(N) \times$ point. Next write the inverse image of the diagonal of $\mathbf{P}^1 \times \mathbf{P}^1$ as $\Delta + \Delta'$ where Δ is the diagonal on $X_0(N) \times X_0(N)$ and Δ' is everything else. A point of Δ' (not over ∞) corresponds to a triple (E, G_1, G_2) where E is an elliptic curve and G_1 and G_2 are two distinct cyclic subgroups of E of order N. Recall that a (general) point of the modular curve $X(N)$ corresponds to a triple (E, p_1, p_2) where E is an elliptic curve and p_1 and p_2 are two distinct points of order N on E that together form a basis for the points of order N on E. Denote by $\langle p_i \rangle$ the subgroup of E generated by p_i. We have a surjective morphism

$$
\begin{array}{ccc}
X(N) & \to & \Delta' \\
(E, p_1, p_2) & \mapsto & (E, \langle p_1 \rangle, \langle p_2 \rangle).
\end{array}
$$

Thus Δ' is irreducible. We conclude that the subspace of

$$H^{1,1}(X_0(N) \times X_0(N), \mathbf{C})$$

spanned by divisors supported on $X_0(N) \times X_0(N) \backslash (X_0(N) \times X_0(N))^0$ is spanned by the 4 classes $p \times X_0(N)$, $X_0(N) \times p$, Δ, and Δ'. But there is one nontrivial relation among these classes: the pull back from $\mathbf{P}^1 \times \mathbf{P}^1$ of the relation that the diagonal is linearly equivalent to $p \times \mathbf{P}^1 + \mathbf{P}^1 \times p$.   Q.E.D.

3

The examples constructed in §2 will in general give components of Hurwitz schemes that parametrize branched covers with rather complicated branching data. Often, the simpler the branching data of the covers being parametrized the simpler the component of the Hurwitz scheme that parametrizes them. This leads one to ask the following question.

QUESTION. Is there a reasonable definition for "not too complicated branching" so that for any component H of a Hurwitz scheme such that H parametrizes covers with "not too complicated branching" it is true that the group of divisors on H modulo either linear, algebraic, or homological equivalence is torsion?

If the answer to this question is yes then possibly one could use the map from a Hurwitz scheme to the moduli space of curves given in (2) to study divisors on $\mathcal{M}_g$. An important goal in this direction would be to provide a new proof of Harer's theorem [**AC, H**]. The least complicated branching one could have is that over each branch point all points except one are unramified and one point has ramification index 2. Even in this case it is not known in general whether the divisor groups are torsion.

4

We end this article with some remarks about the problem of determining when a variety is isomorphic to an irreducible component of some Hurwitz scheme. From (1) we know that a component of a Hurwitz scheme is necessarily an étale finite cover of $(\mathbf{P}^1)^{b-3}\backslash S$. The main theorem of [**DDH**] says that when $b = 4$ any such cover (that is any connected étale finite cover of $\mathbf{P}^1\backslash\{0, 1, \infty\}$) is a component of some Hurwitz scheme. When $b > 4$ we do not yet know whether this is always the case, but we now show how an easy generalization of the construction in [**DDH**] shows that it is true for a large class of examples.

Consider a connected finite étale cover

$$(8) \qquad\qquad \pi : X \to (\mathbf{P}^1)^{b-3}\backslash S.$$

Inside $(\mathbf{P}^1)^{b-3}$ consider a $\mathbf{P}^1$ obtained by letting only one factor vary. Restricting the map (8) to this $\mathbf{P}^1$ we obtain a finite étale cover of $\mathbf{P}^1\backslash B$ where B consists of $0, 1, \infty$ and $b - 4$ other points. There is a unique way of completing this to be a branched cover

$$(9) \qquad\qquad \pi' : C \to \mathbf{P}^1.$$

We say that the cover in (8) has a connected automorphism free slice if and only if for some choice of cover (9) obtained as described, C is connected and has no automorphisms over $\mathbf{P}^1$.

10. PROPOSITION. *If the cover* (8) *has a connected automorphism free slice then* X *is isomorphic to a component of some Hurwitz scheme.*

PROOF. Without loss of generality we may assume the connected automorphism free slice comes by letting only the first factor vary. Think of $(\mathbf{P}^1)^{b-3}$ as $\mathbf{P}^1 \times (\mathbf{P}^1)^{b-4}$, that is a family of $\mathbf{P}^1$'s parametrized by $(\mathbf{P}^1)^{b-4}$. Whether or not a slice is connected or has automorphisms depends on discrete branching data which is constant in such a family. Therefore all slices in this family are connected and automorphism free. For each $p \in X$ we now construct a branched cover $f_p : C_p \to \mathbf{P}^1$. The point p lies in a unique slice from our family of first factor slices. Let $\pi_p : C'_p \to \mathbf{P}^1$ be that slice. Let $p_1, p_2, \ldots, p_{b-4}$ be the points where this base $\mathbf{P}^1$ meets diagonals of $(\mathbf{P}^1)^{b-3}$. Apply Construction (X) with $X := C'_p$ and $S := p \cup \pi_p^{-1}\{0, 1, \infty, p_1, \ldots, p_{b-4}\}$. Rename the resulting X' as C_p and set $f_p = \pi_p \circ g$.

For the same reasons that directly followed Construction $(C \times C)$ we get a commutative diagram

$$X \xrightarrow{\omega} H_{k,b}$$
$$\pi \searrow \qquad \swarrow \pi_{k,b}$$
$$(\mathbf{P}^1)^{b-3} \backslash S ,$$

and to show that ω is an isomorphism onto some component of $H_{k,b}$ we only need to show that it is one-one.

If ω is not one-one there is an isomorphism $\alpha : C_p \to C_q$ over $\mathbf{P}^1$ with $p \neq q$. But clearly $\pi(p) = \pi(q)$ so that $C'_p = C'_q$. An argument similar to that used to construct diagram (6) then shows that there is an automorphism of C'_p over $\mathbf{P}^1$ taking p to q. This contradicts the fact that we have an automorphism free slice. Q.E.D.

In Construction (C) the purpose of $\tilde{C}$ was to get rid of automorphisms. Getting rid of automorphisms in larger dimensional families seems to be more difficult. Proposition 10. shows that components of Hurwitz schemes can be relatively arbitrary varieties. One therefore suspects that it might be rather difficult to prove general theorems about them.

REFERENCES

[AC] E. Arbarello and M. Cornalba, *The Picard groups of the moduli spaces of curves*, Topology **26** (1987), 153–172.

[BK] B. J. Birch and W. Kuyk, eds., *Modular functions of one variable* IV, Lecture Notes in Math., vol. 476, Springer-Verlag, Berlin, 1975.

[DDH] S. Diaz, R. Donagi, and D. Harbater, *Every curve is a Hurwitz space*, Duke Math. J. **59**, (1989), 737–746.

[DH] S. Diaz and J. Harris, *Geometry of the Severi variety*, Trans. Amer. Math. Soc. **309** (1988), 1–34.

[F] W. Fulton, *Hurwitz schemes and the irreducibility of moduli of algebraic curves*, Ann. of Math. **90** (1969), 542–575.

[H] J. Harer, *The second homology group of the mapping class group of an orientable surface*, Invent. Math. **72** (1982), 221–239.

[P.] A. N. Parshin, *Algebraic curves over function fields*. I, Math. USSR-Izv. **2** (1968), 1145–1170.

STEVEN DIAZ, DEPARTMENT OF MATHEMATICS, SYRACUSE UNIVERSITY, SYRACUSE, NY 13244

E-mail address: SPDIAZ AT SUVM.BITNET

RON DONAGI, DEPARTMENT OF MATHEMATICS, UNIVERSITY OF PENNSYLVANIA, PHILADELPHIA, PA 19104

Contemporary Mathematics
Volume 116, 1991

Normal Sheaves of Linear Systems on Curves

LAWRENCE EIN

Let C be a smooth projective curve of genus g. For each positive integer d, we let $C^{(d)}$ be the dth symmetric product of C. Suppose that L is a degree d line bundle on C. Then the projective space $P_L = P(H^0(L)^*)$ is a closed submanifold of $C^{(d)}$. Let N be the normal sheaf of P_L in $C^{(d)}$. In this paper, we investigate the cohomological properties of N.

If $h^0(L) = r + 1$, then P_L is a r-dimensional projective space. We show that N can be reconstructed from the multiplication map $H^0(\mathscr{O}_{P_L}(1)) \otimes H^1(N(-1)) \to H^1(N)$. We believe that the graded module $M = \bigoplus_k H^1(N(k))$ is an interesting invariant associated with the linear system $|L|$. For instance, if C is a general curve, then we show that $M = H^1(N(-1))$ and N is isomorphic to $\rho \mathscr{O}_{P_r} \oplus (r + g - d)\Omega_{P^r}(1)$, where $\rho = g - (r + 1)(r + g - d)$. Let $\mathrm{Pic}^d C$ be the degree d component of the Picard group of C. For each nonnegative integer r, we consider the determinantal subscheme defined by

$$W_d^r = \{L \in \mathrm{Pic}^d C \,|\, h^0(L) \geq r + 1\}.$$

Suppose that L is degree d line bundle on C and $h^0(L) = r + 1 > 0$. Let $[L] \in W_d^0 = W$ be the corresponding point. As an application to our cohomological computation, we show that the Hilbert function of the local ring $\mathscr{O}_{W,[L]}$ depends only on d, g, and r. As another application to our computations, we give a simple proof of a theorem by Martens on the upper bound of the dimension of W_d^r.

1980 *Mathematics Subject Classification* (1985 *Revision*). Primary 14C20, 14C22.
Research partially supported by NSF Grant DMS 89-0443.
This paper is the final form and no version of it will be submitted for publication elsewhere.

We are grateful to M. Green and R. Lazarsfeld for valuable discussions.

$$\S\mathbf{1}$$

Let C be a smooth projective curve of genus g, and for each positive integer d let $C^{(d)}$ be the dth symmetric product of C. Also let $\mathrm{Pic}^d C$ be the degree d component of the Picard group of C and let $W_d^0 = \{L \in \mathrm{Pic}^d C | h^0(L) > 0\}$.

We consider the Abel-Jacobi map,

$$f \colon C^{(d)} \to W_d^0 \subset \mathrm{Pic}^d C.$$

Let L be a line bundle of degree d on C. We assume that $h^0(L) = r+1 \geq 2$. Denote by $[L]$, the corresponding point in $\mathrm{Pic}^d L$. Let $P_L \cong P(H^0(L)^*) = f^{-1}([L])$. We will study the normal sheaf of P_L in $C^{(d)}$. The following result is well known to the experts.

THEOREM 1.1. *Let N be the normal sheaf of P_L in $C^{(d)}$. Then there is the following exact sequence,*

$$(1.1.1) \qquad 0 \to N \to H^1(\mathscr{O}_C) \otimes \mathscr{O}_{P_L} \to H^1(L) \otimes \mathscr{O}_{P_L}(1) \to 0.$$

PROOF. Let $R = P_L \times C$ and let $D \subseteq R$ be the subvariety corresponding to the family of divisors of C parametrized by P_L. We consider the following diagram:

$$C \xleftarrow{\;p\;} R \supseteq D$$
$$\downarrow q$$
$$P_L$$

There is an exact sequence,

$$(1.1.2) \qquad 0 \to \mathscr{O}_R \xrightarrow{\;s\;} p^*L \otimes q^*\mathscr{O}_{P_L}(1) \to \mathscr{O}_D(D) \to 0,$$

where $s \in H^0(p^*L \otimes q^*\mathscr{O}_{P_L}(1)) \cong \mathrm{Hom}(H^0(L), H^0(L))$ is the section corresponding to the identity map. We apply q_* to (1.1.2). We obtain the following exact complex, $0 \to \mathscr{O}_{P_L} \xrightarrow{\;A\;} H^0(L) \otimes \mathscr{O}_{P_L}(1) \to q_*\mathscr{O}_D(D) \xrightarrow{\;B\;} H^1(\mathscr{O}_C) \otimes \mathscr{O}_{P_L} \to H^1(L) \otimes \mathscr{O}_{P_L}(1) \to 0$. We observe that $\mathrm{cok}(A)$ is isomorphic to the tangent sheaf of P_L and $q_*\mathscr{O}_D(D)$ is isomorphic to $T_{C^{(d)}}|_{P_L}$, where $T_{C^{(d)}}$ is the tangent sheaf of $C^{(d)}$. Furthermore, the B is the map induced by df. So we obtain the following exact sequence,

$$0 \to N \to H^1(\mathscr{O}_C) \otimes \mathscr{O}_{P_L} \to H^1(L) \otimes \mathscr{O}_{P_L}(1) \to 0,$$

where N is the normal sheaf of $\dot{P}_L$ in $C^{(d)}$. $\quad\square$

PROPOSITION 1.2. *Let N be the normal sheaf of P_L. Then*
(a) $\mathrm{rank}(N) = d - r$,
(b) $H^1(N(-1)) \cong H^1(L)$.
(c) $H^0(N(-1)) = 0$.

PROOF. This follows from the exact sequence (1.1.1). $\quad\square$

PROPOSITION 1.3. *We $W \subseteq H^0(L)$ be a subspace and let P_W be the corresponding projective subspace of P_L. We consider the multiplication map,*

$$M_W \colon W \otimes H^0(K \otimes L^{-1}) \to H^0(K).$$

Then

(a) $H^0(N|_{P_W}) \cong (\mathrm{cok}(M_W))^*$.

(b) $H^1(N|_{P_W}) \cong (\ker(M_W))^*$. *In particular, $H^0(N)$ is isomorphic to the Zarisiki tangent space of W_d^r at $[L]$, where W_d^r is the subscheme of $\mathrm{Pic}^d C$, defined by, $W_d^r = \{L \in \mathrm{Pic}^d C | h^0(L) \geq r+1\}$.*

(c) *Let N_W be the normal sheaf of P_W in $C^{(d)}$. Then $N_W \cong N|_{P_W} \oplus m\mathscr{O}_{P_W}(1)$ where $m = \dim P_L - \dim P_W$.*

PROOF. We restrict the exact sequence (1.1.1) to P_W. We observe that $H^0(\mathscr{O}_{P_W}(1)) \cong W^*$, and obtain an exact sequence.

$$0 \to H^0(N|_{P_W}) \to H^1(\mathscr{O}_C) \xrightarrow{A} H^1(L) \otimes W^* \to H^1(N|_{P_W}) \to 0.$$

We check that $A = M_W^*$. It follows that $H^0(N|_{P_W}) = (\mathrm{cok}(M_W))^*$ and $H^1(N|_{P_W}) = (\ker(M_W))^*$. We observe that the normal sheaf of P_W in P_L is isomorphic to $m\mathscr{O}_{P_W}(1)$. By (1.1.1)*, we see $H^1(N^*(1)|_{P_W}) = 0$. So the exact sequence of conormal sheaves,

$$0 \to N^*(1)|_{P_W} \to N_W(1) \to m\mathscr{O}_{P_W} \to 0,$$

is split exact. It follows that

$$N_W \cong N|_{P_W} \oplus m\mathscr{O}_{P_W}(1). \quad \square$$

COROLLARY 1.4. *$P_W \subseteq P_L$ is the complete linear system if and only if $H^0(N_W(-1)) = 0$.*

PROOF. This follows from 1.2(C) and 1.3(C). $\square$

In [B], Bloch constructs the semiregularity map, in studying deformations of submanifolds. In our case, P_L is a codimension $d-r$ closed submanifold of $C^{(d)}$. So we obtain a semiregularity map:

$$S_L \colon H^1(N) \to H^{d-r+1}(\Omega_{C^{(d)}}^{d-r+1}).$$

PROPOSITION 1.5. *If S_L is injective, then W_d^r is smooth at $[L]$.*

PROOF. If S_L is injective, then the Hilbert scheme of $C^{(d)}$ is smooth at $[P_L]$ by Bloch's theorem ([B]). Let $n = h^0(N)$. Then we can find a n-dimensional open set U of the Hilbert scheme of $C^{(d)}$ containing $[P_L]$ parametrizing r-planes in $C^{(d)}$. The Abel-Jacobi map induces a map

$$h \colon U \to \mathrm{Pic}^d C.$$

By (1.1.1), the differential of h at P_L,

$$dh\colon H^0(N) \to H^1(\mathscr{O}_C)$$

is injective. We conclude that $\dim h(U) = n$. It follows that the dimension of W_d^r at $[L]$ is greater or equal to n. By 1.3(b), the dimension of the Zariski tangent space of W_d^r at $[L]$ is equal to n. So we conclude that W_d^r is smooth that $[L]$.   □

§2

In this section, we study the cohomological properties of N as a locally free sheaf on P^r.

PROPOSITION 2.1. (a) *There is a locally free sheaf* N_1 *on* P_L *such that* $N \cong N_1 \oplus H^0(N) \otimes \mathscr{O}_{P_L}$.
   (b) $H^i(N(j)) = 0$ *for* $2 \le i \le r-1$ *and arbitrary* j.
   (c) $H^r(N(j)) = 0$ *if* $r \ge 2$ *and* $j \ge -r$.

PROOF. (a) Consider the following diagram

$$
\begin{array}{ccc}
0 & & 0 \\
\downarrow & & \downarrow \\
H^0(N) \otimes \mathscr{O}_{P_L} & = & H^0(N) \otimes \mathscr{O}_{P_L} \\
\downarrow & & \downarrow \\
0 \to N & \to & H^1(\mathscr{O}_C) \otimes \mathscr{O}_{P_L}.
\end{array}
$$

Let p be a projection map from $H^1(\mathscr{O}_C)$ to $H^0(N)$. Then $p \circ h$ induces a projection map from N to $H^0(N) \otimes \mathscr{O}_{P_L}$. Set $N_1 = \ker(p \circ h)$. We see that $N \cong N_1 \oplus H^0(N) \otimes \mathscr{O}_{P_L}$.
   (b) and (c) follow from the exact sequence (1.1.1).   □

PROPOSITION 2.2. *We can reconstruct the locally free sheaf* N *from the multiplication map,* $B\colon H^0(\mathscr{O}_{P_L}(1)) \otimes H^1(N(-1)) \to H^1(N)$.

PROOF. Let N_1 be as in 2.1(a). We observe that $H^1(N_1(j)) \cong H^1(N(j))$ for $j \ge -r = -\dim P_L$. Using the spectral sequence of Beilinson [OSS, p. 240], and 2.1, we obtain an exact sequence

$$(2.2.1) \qquad 0 \to N_1 \to H^1(N_1(-1)) \otimes \Omega_{P_L}(1) \xrightarrow{A} H^1(N_1) \otimes \mathscr{O}_{P_L} \to 0.$$

But A is the composition of the following maps

$$H^1(N_1(-1)) \otimes \Omega_{P_L}(1) \xrightarrow{1 \otimes F} H^1(N_1(-1)) \otimes H^0(\mathscr{O}_{P_L}(1)) \otimes \mathscr{O}_{P_L} \to H^1(N_1) \otimes \mathscr{O}_{P_L}$$

where F is the natural map from $\Omega_{P_L}(1) \to H^0(\mathscr{O}_{P_L}(1)) \otimes \mathscr{O}_{P_L}$. So N_1 can be reconstructed from B. But N is just simply $N_1 \oplus (d - r - \operatorname{rank} N_1)\mathscr{O}_{P_L}$.   □

THEOREM 2.3. *Let* $\rho = g - (r+1)(g+r-d)$ *be the Brill-Noether number.*
(a) *If* $H^1(N) = 0$, *then* $N \cong \rho \mathscr{O}_{P_L} \oplus H^1(L) \otimes \Omega_{P_L}(1)$.
(b) *If* C *is a general curve, then* $H^1(N) = 0$.

PROOF. By 2.2.1 and 1.2(b), $N_1 \cong H^1(L) \otimes \Omega_{P_L}(1)$. Since $H^1(N) = 0$, the multiplication map

$$M: H^0(L) \otimes H^0(K \otimes L^{-1}) \to H^0(K)$$

is injective by 1.3(b). So $h^0(N) = \rho$ and $N \cong \rho \mathscr{O}_{P_L} \oplus H^1(L) \otimes \Omega_{P_L}(1)$ by 2.1(a).
(b) By a theorem of Gieseker [**G**], the multiplication map M is injective for a general curve C. It follows that $H^1(N) = 0$. $\square$

Let F be a coherent sheaf on P^r. We recall the notion of Castelnuovo-Mumford regularity. We say that F is k-regular, if $H^i(F(k-i)) = 0$ for $i > 0$ [**M**, p. 100]. It follows that $H^i(F(t-i)) = 0$ for $i > 0$ and $t \geq k$.

PROPOSITION 2.4. *Let* A *and* B *be the two finite dimensional vector spaces. Let* $a = \dim A$ *and* $b = \dim B$. *We suppose that there is the following exact sequence on* P^r.

(2.4.1) $$0 \to B \otimes \mathscr{O}_{P^r}(-1) \to A \otimes \mathscr{O}_{P^r} \to E \to 0,$$

where E *is a locally free sheaf on* P^r.
(a) E *is 0-regular.*
(b) *For each positive integer* p, $S^p E$, *the* pth *symmetric product of* E, *is 0-regular.*
(c) *The natural multiplication map* $S^p A = S^p H^0(E) \to H^0(S^p E)$ *is surjective.*
(d) E^* *is* b-*regular.*

PROOF.
(a) This follows from an easy computation using (2.4.1).
(b) and (c) There is an Eagon-Northcott complex,

(2.4.2) $$0 \to S^{p-b} A \otimes \Lambda^b B \otimes \mathscr{O}_{P^r}(-b) \to \cdots \to S^{p-1} A \otimes B \otimes \mathscr{O}_{P^r}(-1)$$
$$\to S^p A \otimes \mathscr{O}_{P^r} \to S^p E \to 0.$$

Using (2.4.2), we can check that $S^p E$ is 0-regular and $S^p A$ maps onto $H^0(S^p E)$.
(d) Let $F = A^* \otimes \mathscr{O}_{P^r}$ and $G = B^* \otimes \mathscr{O}_{P^r}(1)$. Then we have an exact sequence

(2.4.3) $$0 \to E^* \to F \to G \to 0.$$

Using (2.4.3), we obtain an Eagon-Northcott complex,

(2.4.4) $$0 \to \Lambda^a F \otimes (S^{a-b-1} G)^* \to \cdots \to \Lambda^{b+1} F \to E^* \otimes \Lambda^b G \to 0.$$

Observe that $E^* \otimes \Lambda^b G \cong E^*(b)$. Using (2.4.4), we can check that E^* is b-regular. $\square$

REMARK. The 0-regularity of $S^p E$ would also follow from a Lemma of R. Lazarsfeld.

PROPOSITION 2.5. *Let E be a locally free sheaf on P^r as defined in (2.4.1). Then* $\operatorname{rank} E \geq r$.

PROOF. Let $h: P(B^* \otimes \mathscr{O}_{P^r}(1)) \to P(A^*)$ be the natural projection map. $h^* \mathscr{O}_{P(A^*)}(1)$ is the tautological line bundle U of $P(B^* \otimes \mathscr{O}_{P^r}(1))$. Since $B^* \otimes \mathscr{O}_{P^r}(1))$ is an ample vector bundle, U is also ample. We conclude that h is a finite morphism. So $\dim P(A^*) = a - 1 \geq \dim P(B^* \otimes \mathscr{O}(1)) = b - 1 + r$. It follows that $\operatorname{rank} E = a - b \geq r$. $\square$

THEOREM 2.6. (a) N^* *and* $S^p N^*$ *are 0-regular.*
(b) N *is* $(r + g - d)$-*regular.*
(c) $H^1(N(k)) = 0$ *for* $k \geq r + g - d - 1$.

PROOF. This follows from the exact sequence (1.1.1) and Proposition 2.5. $\square$

THEOREM 2.7. *Let $\widehat{C}^{(d)}$ be the formal completion of $C^{(d)}$ along P_L.*
(a) $H^i(\widehat{C}^{(d)}), \widehat{\mathscr{O}}_{C^{(d)}}) = 0$ *for* $i > 0$.
(b) *Let* $f: C^{(d)} \to \operatorname{Pic}^d C$ *be the Abel-Jacobi map. Then* $R^i f_*(\mathscr{O}_{C^{(d)}}) = 0$ *for* $i > 0$.
(c) $H^i(\widehat{C}^{(d)}; \widehat{\omega}_{C^{(d)}}) = 0$ *for* $i \geq \operatorname{Max}(1, d + 1 - g)$.
(d) $R^i f_*(\omega_{C^{(d)}}) = 0$ *for* $i \geq \operatorname{Max}(1, d + 1 - g)$.

PROOF. Let I be the ideal sheaf of P_L and let kP_L be the closed subscheme of $C^{(d)}$ defined by the ideal I^k. We have an exact sequence,

$$(2.7.1) \qquad\qquad 0 \to S^k N^* \to \mathscr{O}_{(k+1)P_L} \to \mathscr{O}_{kP_L} \to 0.$$

By 0-regularity of $S^k N^*$ and induction, we see $H^i(\mathscr{O}_{(k+1)P_L}) = 0$ for $i > 0$. Hence $H^i(\widehat{C}^{(d)}, \widehat{\mathscr{O}}_{C^{(d)}}) = 0$ for $i > 0$. $R^i f_*(\mathscr{O}_{C^{(d)}}) = 0$ for $i > 0$ by the formal functions theorem. Observe that

$$\omega_{C^{(d)}}|_{P_L} \cong \det N^* \otimes \mathscr{O}_{P_L}(-r - 1) \cong \mathscr{O}_{P_L}(g - d - 1)$$

by the adjunction formula. We see that $H^i(S^k N^* \otimes \mathscr{O}(g - d - 1)) = 0$ for $i \geq \operatorname{Max}(1, d + 1 - g)$ by the 0-regularity of $S^k N^*$. Similarly we conclude that $H^i(\widehat{C}^{(d)}, \widehat{\omega}_{C^{(d)}}) = 0$ for $i \geq \operatorname{Max}(1, d + 1 - g)$ and $R^i f_*(\omega_{C^{(d)}}) = 0$ for $i \geq \operatorname{Max}(1, d + 1 - g)$. $\square$

REMARK 2.8. 2.7(c) is a theorem of Kempf ([K]). There he shows that $W_d^0(d \leq g - 1)$ has rational singularity.

THEOREM 2.9 (Martens). *Let $W_d^r = \{L \in \operatorname{Pic}^d C | h^0(L) \geq r + 1\}$. Suppose that $h^0(L) = r + 1$ and $h^1(L) \neq 0$.*
(a) $\dim W_d^r \leq d - 2r$.

(b) *Assume that $d \leq g - 1$ and $r \geq 1$. If $\dim W_d^r = d - 2r$, then C is a hyperelliptic curve.*

PROOF. (a) By 2.5, $\operatorname{rank} N_1 \geq r$. We see that $d - r = \operatorname{rank} N = h^0(N) + \operatorname{rank} N_1$. We conclude that $h^0(N) \leq d - 2r$. Since $H^0(N)$ is the Zariski tangent space of W_d^r at $[L]$, we conclude that the dimension of W_d^r is less than or equal to $d - 2r$ at the point $[L]$.

(b) If $d - 2r = \dim W_d^r = 0$, then C is hyperelliptic by Clifford's theorem. So we assume that $\dim W_d^r = d - 2r \geq 1$. Let D be a general degree $r - 1$ effecting divisor on C. Then $W = H^0(L(-D))$ defines a line P_W in P_L. Consider the exact sequence

$$0 \to L^*(D) \to W \otimes \mathscr{O}_C \to L(-D) \to 0,$$

and

$$0 \to L^{-2} \otimes K_C \to W \otimes K \otimes L^*(D) \to K_C \to 0.$$

Let M_W be the multiplication map $M_W \colon W \otimes H^0(K \otimes L^*(D)) \cong W \otimes H^0(K \otimes L^*) \to H^0(K_C)$. By Proposition 1.3, $\operatorname{Ker}(M_W) = h^0(L^{-2} \otimes K_C) = h^1(N|_{P_W}) \geq -\chi(N|_{P_W}) = g - d$. By the Riemann-Roch theorem, we conclude that $h^0(L^2) \geq d + 1$. By Clifford's theorem, either $L^2 = K_C$ or C is hyperelliptic since $\dim W_d^r \geq 1$, we may assume that $L^2 \neq K_C$. It follows from Theorem 2.7 that C is a hyperelliptic curve. $\square$

REMARK. Theorem 2.9 is a theorem of Martens. See [ACGH, p. 191] for another proof and further developments.

§3

Let A, B and E be as in 2.4.1. Let $X = V(E)$, $Z = V(A \otimes \mathscr{O}_{p^n})$ and $W = \Psi(A)$. Consider the diagram:

$$
\begin{array}{ccc}
X & \xrightarrow{\;h\;} & Y \\
\cap| & & \cap| \\
Z & \to & V(A) = W
\end{array}
$$

where p and q are the natural projection maps. Let $h = q|_X$ and $Y = h(X)$.

PROPOSITION 3.1.
(a) *If $\dim X \leq a$, then $h \colon X \to Y$ is a birational map.*
(b) $h_* \mathscr{O}_X = \mathscr{O}_Y$ and $R^i h_* \mathscr{O}_X = 0$, for $i > 0$.
(c) $H^0(\mathscr{O}_Y)$ is isomorphic to the graded ring $\bigoplus H^0(S^k E)$.

PROOF. Consider the composition map:

$$S \colon p^* B \otimes \mathscr{O}_{p^n}(-1) \to A \otimes \mathscr{O}_Z \to \mathscr{O}_Z.$$

We observe that X is the zero scheme of S. We have the following Koszul complex

$$(3.1.1) \quad 0 \to \Lambda^b(B \otimes p^* \mathscr{O}_{p^n}(-1)) \to \cdots \to p^*(B \otimes \mathscr{O}_{p^n}(-1)) \to \mathscr{O}_Z \to \mathscr{O}_X \to 0.$$

If $k \geq n+1$, $R^n q_*(\Lambda^k(B \otimes p^* \mathcal{O}_{p^n}(-1))) \cong \Lambda^k B \otimes H^n(\mathcal{O}_{p^n}(-k)) \otimes \mathcal{O}_W$. We apply q_* to 3.1.1. From the spectral sequence associated with the complex, we obtain the following Eagon-Northcott complex,

$$0 \to \Lambda^b B \otimes H^n(\mathcal{O}_{p^n}(-b)) \otimes \mathcal{O}_W \to \cdots \to \Lambda^{n+1} B \otimes H^n(\mathcal{O}_{p^n}(-n-1)) \otimes \mathcal{O}_W$$
$$\to \mathcal{O}_W \to q_* \mathcal{O}_X \to 0.$$

We see that $q_* \mathcal{O}_X = \mathcal{O}_Y$. Since cohomological dimension of $\mathcal{O}_Y$ is less than or equal to $(b-n) = a - \dim X$. We see that $\dim Y = \dim X$. Hence $h\colon X \to Y$ is birational. Using (3.1.1), we also see that $R^i h_* \mathcal{O}_X = 0$ for $i > 0$. Also $H^0(\mathcal{O}_X) \cong H^0(\mathcal{O}_Y)$. Since $X = V(E)$, $H^0(\mathcal{O}_X)$ is the graded ring $\bigoplus H^0(S^k E)$. $\square$

Let $W = W_d^0$ and $M \subseteq \mathcal{O}_W$ be the maximal ideal defining the point $[L]$. Let $I \subseteq \mathcal{O}_{C^{(d)}}$ be the ideal sheaf of P_L. Let kP_L be the subscheme defined I^k. Consider the Abel-Jacobi map, $f\colon C^{(d)} \to W$.

By the formal functions theorem we see that,

$$(3.2.1) \qquad \varprojlim H^0(\mathcal{O}_{kP_L}) \cong \varprojlim \mathcal{O}_W / M^k.$$

THEOREM 3.2 (Kempf). *Assume that $d \leq g-1$. We consider the natural map, $h\colon V(N^*) \to V(H^0(K))$. Let $Y = V(N^*)$. Then Y is the tangent cone of W at $[L]$.*

PROOF. Let T be the tangent cone of W_d at $[L]$. By blowing up, one can see that the reduced affine variety Y is a closed subscheme of T. We will show that the natural surjection $H^0(\mathcal{O}_T) \to H^0(\mathcal{O}_Y)$ is an isomorphism. This will imply the affine schemes T and Y are the same. For each nonnegative integer, we consider the following diagram:

$$
\begin{array}{ccccccccc}
0 & \longrightarrow & M^k/(M^{k+1}) & \longrightarrow & \mathcal{O}_W(M^{k+1}) & \longrightarrow & \mathcal{O}_W/(M^k) & \longrightarrow & 0 \\
 & & \downarrow & & \downarrow {\scriptstyle s_{k+1}} & & \downarrow {\scriptstyle t_{k+1}} & & \downarrow {\scriptstyle t_k} \\
0 & \longrightarrow & H^0(S^k N^*) & \longrightarrow & H^0(\mathcal{O}_{k+1 P_L}) & \longrightarrow & H^0(\mathcal{O}_{k P_L}) & \longrightarrow & 0.
\end{array}
$$

Since $H^1(S^k N^*) = 0$ by 2.6(a), the bottom row is exact. $H^0(\mathcal{O}_Y)$ is the graded ring $\bigoplus H^0(S^k N^*)$ by 3.1(c). We see that s_{k+1} is surjective for $k \geq 0$. If s_{k+1} is not injective, then t_{k+1} will fail to be injective. Since s_{k+2} is surjective, we see that $\ker(t_{k+2})$ maps onto $\ker(t_{k+1})$ by the snake lemma. In this way, we will be able to construct an element in the kernel of the map $t\colon \varprojlim \mathcal{O}_W/(M^k) \to \varprojlim H^0(\mathcal{O}_{kP_L})$. But this contradicts (3.2.1). We conclude that s_{k+1} is an isomorphism for each k. It follows that $H^0(\mathcal{O}_T) \cong H^0(\mathcal{O}_Y)$ and $Y = T$. $\square$

THEOREM 3.3. *Let M be the maximal ideal in W corresponding to the* *point* $[L]$. *Then*

(a) $\dim M^k/(M^{k+1}) = H^0(S^k N^*)$.

(b) *Hilbert function of the local ring* $\mathcal{O}_{W,[L]}$ *depends only on* d, g, *and* r.

PROOF. (a) In the proof of 3.2, we have established the isomorphism between $M^k/(M^{k+1})$ and $H^0(S^k N^*)$.

(b) Since $S^k N^*$ is 0-regular, $\chi(S^k N) = h^0(S^k N^*)$ depends only on the Chern classes of N by the Riemann-Roch formula. By the exact sequence (1.1.1), we see that the Chern classes of N depends only on d, g, and r.

The following are some examples.

(3.4.1). Assume that $\deg L = g - 1$, $r = 1$, and L is generated by its sections. We consider the exact sequence

$$0 \to L^{-2} \otimes K \to H^0(L) \otimes K \otimes L^{-1} \to K \to 0.$$

By Proposition 1.3, there are the following two possibilities.

(a) $L^2 = K$. Then $h^1(N) = 1$ and $N^* \cong (g-3)\mathcal{O}_{p^1} \oplus \mathcal{O}_{p^1}(2)$. One can see that the tangent cone T is a cone over a conic in P^2.

(b) $L^2 \neq K$. Then $h^1(N) = 0$ and $N^* = g - 4\mathcal{O}_{p^1} \oplus 2\mathcal{O}_{p^1}(1)$. One checks in this case that T is a cone over a quadric surface in P^3.

(3.4.2) If $r = 1$ and $M_{H^0(L)}$ is injective, then $N^* \cong (2d-g-2)\mathcal{O}_{p^1} \oplus g + 1 - d\mathcal{O}_{p^1}(1)$. Then the tangent cone is a cone over $P^1 \times P^{g-d}$.

(3.4.3) Assume that $d \leq g-1$ and $M_{H^0(L)}$ is injective. Then $N^* \cong \rho\mathcal{O}_{p^r} \oplus (g+r-d)T_{p^r}(-1)$, where $\rho = g-(r+1)(r+g-d)$ and T_{p^r} is the tangent sheaf of p^r by Theorem (2.3). Observe that the natural image of $(g+r-dT_{p^r}(-1))$ is in the affine space of dimension $(g+r-d)(r+1)$ corresponds to those $(r+1) \times (g+r-d)$ matrices of rank less than or equal to r. The tangent cone T is a cone over this determinantal variety.

§4

In this section we study a spacial example. In this case, we will see that there are closed relations between the sheaf N and the properties of the linear system $|L|$. Let C be a hyperelliptic curve of genus $g(g \geq 4)$ and $p: C \to P^1$ be the two to one covering map. Let $L = p^*\mathcal{O}_{p^1}(2)$. Then $\deg L = 4$ and $h^0(L) = 3$. Let N be the normal sheaf of $P_L = P(H^0(L)^*)$ in $C^{(4)}$.

THEOREM 4.1. *Let* C, L, *and* N *be as above. Then*

(a) N *is a stable* rank 2 *vector bundle on* P^2.

(b) $h^1(N(t)) \neq 0$ *for* $-1 \leq t \leq g - 4$.

(c) *Let* $h: C \to P(H^0(L)) = P_L^*$ *be the natural map. Then* $D = h(C)$ *is a conic. If* $x \in P_L^*$. *We denote by* 1_x *the corresponding line in* P_L. *Then* $h^0(N|_{1_x}) \neq 0$ *if and only if* $x \neq D$.

PROOF. (a) $\Lambda^2 N^* \cong \mathscr{O}_{p^2}(g-2)$. The perfect pairing $N^* \times N^* \to \Lambda^2 N^*$ implies that $N(g-2) \cong N^*$. It follows from the definition of stability that N is stable if and only if $h^0(N(t)) = 0$ for $t \leq (g-2)/2$. Using (1.1.1) we see that $h^0(N(g-3)) = h^0(N^*(-1)) = 0$. We conclude that N is a stable vector bundle.

(b) By dualizing (1.1.1), we obtain

$$(4.1.1) \quad 0 \to H^0(K \otimes L^{-1}) \otimes \mathscr{O}_{p^2}(-1) \to H^0(K) \otimes \mathscr{O}_{p^2} \to N(g-2) \to 0.$$

Using (4.1.1), a simple calculation shows that $h^1(N(g-4)) \neq 0$ and $h^1(N(k)) = 0$, for $k \geq g-3$. By Serre's duality, we see that $h^1(N(k)) = 0$ for $k \leq 2$. On the other hand, suppose that $h^1(N(k)) = 0$ for some k satisfy $0 \leq k \leq g-4$. Since $H^2(N(k-1)) = 0$, we see that N is $(k+1)$-regular. This implies that $h^1(N(g-4)) = 0$ which is a contradiction. We conclude that $h^1(N(k)) \neq 0$ for $-1 \leq k \leq g-4$.

(c) Let $x \in D \cong P^1$. Let $p^{-1}(x) = \{y_1, y_2\}$ be the inverse image of x. The line 1_x corresponds to the pencil $|L(-y_1 - y_2)|$. Using Proposition 1.3, we see that $h^0(N|_{1_x}) = 1$, if $z \notin D$, then l_z corresponds to a base points free pencil. Again using 1.3 and the base points free pencil trick, we can check that $h^0(N|_{l_z}) = 0$. □

REMARK. The computation in 4.1(b) shows that the vanishing theorem in 2.6(c) gives the optimal bound.

REFERENCES

[ACGH] E. Arbarello, M. Cornalba, P. Griffiths, and J. Harris, *Geometry of Algebraic Curves*, Vol. I, Springer-Verlag, Berlin-Heidelberg-New York, 1984.

[B] S. Bloch, *Semiregularity and de Rham cohomology*, Invent. Math. **17** (1972) 51–66.

[E] L. Ein, *An analogue of Max Noether's theorem*, Duke Math. J. **52** (1985), 689–706.

[G] D. Gieseker, *Stable curves and special divisors, Petri's conjecture*, Invent Math. **66** (1982), 251–275.

[K] G. Kempf, *On the geometry of a theorem of Riemann singularities*, Ann. of Math. **98** (1973), 178–185.

[M] D. Mumford, *Lectures on curves on an algebraic surface*, Ann. of Math. Stud., vol. 59, Princeton Univ. Press, Princeton, 1966.

[OSS] C. Okonek, M. Schneider, and H. Spindles, *Vector bundles on complex projective space*, Progr. Math., Birkhauser, Boston, 1980.

DEPARTMENT OF MATHEMATICS, UNIVERSITY OF ILLINOIS AT CHICAGO, M/C 249, BOX 4348, CHICAGO, ILLINOIS 60608

Contemporary Mathematics
Volume 116, 1991

An Overview of the Geometry
of Algebraic Fermi Curves

D. GIESEKER, H. KNÖRRER, AND E. TRUBOWITZ

1. Introduction

Let $\Gamma \subset R^d$, $d \leq 3$, be a lattice of maximal rank and q a real valued function belonging to $L^2(R^d/\Gamma)$. The conventional mathematical theory (for example, [RS]) of the periodic Schrödinger operator $-\Delta + q(x)$ consists in showing by analytic means that its spectrum is absolutely continuous and verifying that the corresponding Bloch solutions are complete. Recall that a Bloch solution ψ with crystal momentum k in R^d satisfies:

$$(1) \qquad (-\Delta + q(x))\psi = \lambda\psi,$$

for some real λ, and

$$(2) \qquad \psi(x + \gamma) = e^{i\langle k, \gamma\rangle}\psi(x),$$

for all γ in Γ. Essentially, ψ is a plane wave $e^{i\langle k, x\rangle}$ modulated by the periodic function $e^{-i\langle k, x\rangle}\psi(x)$.

For each k in R^d, (1) and (2) define a selfadjoint boundary value problem which has a discrete spectrum usually denoted by

$$E_1(k) \leq E_2(k) \leq E_3(k) \leq \cdots.$$

The eigenvalue $E_n(k)$, $n \geq 1$, defines a function of k called the nth band function. It is continuous and periodic with respect to the lattice

$$\Gamma^{\#} = \{b \in R^d | \langle\gamma, b\rangle \in 2\pi Z \text{ for all } \gamma \text{ in } \Gamma\},$$

dual to Γ. The set of all band functions is the energy-crystal momentum dispersion relation.

1980 *Mathematics Subject Classification* (1985 *Revision*). Primary 14H10.
Research of the first author was partially supported by NSF Grant DMS86-03175.
The detailed version of this paper has been submitted for publication elsewhere.

In the independent electron approximation of solid state physics (see [AM]), ions are fixed at the points of Γ and each electron moves in the common field $-\nabla q$. For a given electron density the (Fermionic) ground state energy of the system (at temperature zero) is the Fermi energy. The boundary of the region in k-space parametrizing the single particle wave functions, that is the Bloch solutions of (1), that form the many body ground state is called the (physical) Fermi surface. Generally, for an arbitrary energy λ the level surface

$$\{k \in R^d | \text{ there is an } n \geq 1 \text{ with } E_n(k) = \lambda\},$$

is called the Fermi surface over λ. There is a large literature in physics devoted to band functions and Fermi surfaces. It is primarily phenomenological.

During the mid-nineteen seventies a rich structure emerged in one dimension. The successful attempts to solve the Korteweg-de Vries equation explicitly made it natural to attach to each operator $-\frac{d^2}{dx^2} + q(x)$, $q \in L^2(R^1/\Gamma)$ a complex analytic curve; specifically, the plane curve:

$$B(q) := \{(\xi, \lambda) \in \mathbb{C}^* \times \mathbb{C}| \text{ there is a } \psi \text{ in } H^2_{\text{loc}}(R^1)$$

$$\text{satisfying } \left(-\frac{d^2}{dx^2} + q(x)\right)\psi = \lambda\psi \text{ and}$$

$$\psi(x + \gamma) = \xi^\gamma \psi(x) \text{ for all } \gamma \in \Gamma\}.$$

Clearly, for each $n \geq 1$, the graph $\{(e^{ik}, E_n(k))|k \in R\}$ of $E_n(k)$ in $S^1 \times R$ is a subset of $B(q)$. Conversely, the intersection $B(q) \cap (S^1 \times R)$ coincides with these graphs. The curve $B(q)$ contains all points that can possibly be reached by analytic continuation of any band function. Loosely speaking, it is the complete complex energy-crystal momentum dispersion relation. One can easily show (see for example [MT]) that projection onto the λ-plane presents $B(q)$ as a (transcendental) hyperelliptic curve. Thus, for generic λ, the "Fermi surface" over λ consists of two points.

The correspondence between $-\frac{d^2}{dx^2} + q(x)$ and $B(q)$ is very useful. For instance, the Jacobian variety of $B(q)$ parametrizes the space of all potentials in $L^2(R^1/\Gamma)$ with the same spectrum as q.

One would hope that there is an equally interesting analogue of this correspondence in two or more dimensions. To this end, let us introduce for each operator $-\Delta + q(x)$, $q \in L^2(R^d/\Gamma)$, the *Bloch variety*

$$B(q) := \{(\xi_1, \ldots, \xi_d; \lambda) \in (\mathbb{C}^*)^d \times \mathbb{C}| \text{ there is a nontrivial } \psi$$

$$\text{in } H^2_{\text{loc}}(R^d) \text{ satisfying } (-\Delta + q(x))\psi = \lambda\psi$$

$$\text{and } \psi(x + \gamma) = \xi_1^{\gamma_1}\zeta_2^{\gamma_2}\cdots\zeta_d^{\gamma_d}\psi(x) \text{ for all } \gamma \in \Gamma\}.$$

Again, $B(q)$ contains all points in $(\mathbb{C}^*)^d \times \mathbb{C}$ that can be reached by analytic continuation of any band function. It can be shown that $B(q)$ is a

d-dimensional complex analytic variety. For the construction of an analytic equation defining $B(q)$ see [**KT**].

We want to investigate the geometry of $B(q)$ and determine what information it contains about the operator $-\Delta + q$ and its spectrum. However, the fact that $B(q)$ is now a transcendental complex surface or respectively three-fold makes everything inherently more complicated than the one-dimensional situation. The geometry of surfaces (resp. threefolds) is far more intricate than that of curves, and furthermore, since the varieties are *not* algebraic, there are analytical difficulties. To isolate the purely geometrical aspects of the problem we shall discretize $-\Delta + q$ and study the corresponding algebraic Bloch variety. Furthermore, we shall confine the present discussion to two dimensions. The discretized model is formulated in §2 where some of its general properties are discussed.

Most of the time it is not possible to generalize from the case of ordinary differential to partial differential operators. The extremely rigid structure in one dimension seems to evaporate. For us the main point of our work is that the geometry provides a context in which almost everything has a controllable multidimensional analogue. For instance, the branch points of a hyperelliptic curve determine it uniquely since the monodromy is specified. From this remark it is immediate in one dimension that the density of states determines the Bloch variety. With considerably more effort the same can be shown for the generic discrete periodic Schrödinger operator in two dimensions.

The integrated density of states is a physically measurable function $\rho(\lambda)$. In §6 we give its definition and show that $d\rho/d\lambda$ can be realized as a period integral of an explicit relative differential form for the family of complexified Fermi curves. The argument that $\rho(\lambda)$ generically determines the Bloch variety $B(q)$ then uses Torelli's theorem for algebraic curves and Deligne's theorem of the fixed part which implies that an algebraic family of Hodge structures with absolutely irreducible monodromy is determined by its monodromy. This argument is sketched in §§7 and 8. The application of the theorem of the fixed part is based on a detailed analysis of the topology and monodromy of the family of complexified Fermi curves which we describe in §§4 and 5. We also indicate how this topological information about the family of Fermi curves for a generic potential q is obtained by degeneration to constant and separable potentials. To prepare this, and as an introduction to the general case, we discuss the situation of constant and separable potentials in §3.

In §2 we also demonstrate that generically $B(q)$ determines q (up to the obvious symmetries). Putting all this together one learns that generically any germ of the density of states determines the periodic potential and thus the independent electron approximation.

This article is intended as a relatively brief overview of our work in this subject. We shall give a full exposition in another publication.

2. General properties

Let a and b be distinct odd primes. Denote, as usual, by

$$L^2 = L^2(Z^2/aZ \oplus bZ)$$

the (finite dimensional) Hilbert space of complex-valued functions on the discrete torus $Z^2/aZ \oplus bZ$ with inner product

$$\langle \varphi, \psi \rangle = \frac{1}{ab} \sum_{\substack{1 \le m \le a \\ 1 \le n \le b}} \varphi(m, n)\overline{\psi}(m, n).$$

The elements of L^2 may of course be identified with periodic functions on Z^2. The discrete Laplace operator Δ is the bounded linear operator on $l^2(Z^2)$ defined by

$$\Delta\psi(m, n) = \psi(m + 1, n) + \psi(m - 1, n) + \psi(m, n + 1) + \psi(m, n - 1).$$

Clearly, it may also be applied to any complex valued function on Z^2.

For every V in $L^2(Z^2/aZ \oplus bZ)$ we may consider the discrete periodic Schrödinger operator $-\Delta + V$ and define its corresponding (affine) Bloch variety by

$$B(V) := \{(\xi_1, \xi_2, \lambda) \in \mathbb{C}^* \times \mathbb{C}^* \times \mathbb{C}| \text{ there is a complex-valued}$$

$$\text{function } \psi \text{ on } Z^2 \text{ satisfying } (-\Delta + V)\psi = \lambda\psi \text{ and}$$

$$\psi(m + a, n) = \xi_1\psi(m, n), \psi(m, n + b) = \xi_2\psi(m, n) \text{ for all}$$

$$(m, n) \text{ in } Z^2\}.$$

One may write the spectral problem defining $B(V)$ as an $ab \times ab$-matrix acting on the vector $(\psi(m, n), 1 \le m \le a, 1 \le n \le b)$ in $\mathbb{C}^{ab}$. The determinant of this matrix is a polynomial $P(\xi_1, \xi_2, \lambda; V)$ in $\xi_1, \xi_1^{-1}, \xi_2, \xi_2^{-1}$ and λ of degree ab whose vanishing displays $B(V)$ as an algebraic hypersurface in $\mathbb{C}^* \times \mathbb{C}^* \times \mathbb{C}$. Let π_{aff} be the projection of $B(V)$ to $\mathbb{C}$ given by

$$
\begin{array}{ccc}
(\xi_1, \xi_2, \lambda) & \in & B(V) \\
\downarrow & & \downarrow \pi_{\text{aff}} \\
\lambda & \in & \mathbb{C}.
\end{array}
$$

The fibers $F_{\text{aff}, \lambda} := \pi_{\text{aff}}^{-1}(\lambda)$ are called (affine) Fermi curves. Furthermore, let i be the involution on $\mathbb{C}^* \times \mathbb{C}^* \times \mathbb{C}$ that sends (ξ_1, ξ_2, λ) to $(1/\xi_1, 1/\xi_2, \lambda)$. We have

$$P(\xi_1, \xi_2, \lambda; V) = P\left(\frac{1}{\xi_1}, \frac{1}{\xi_2}, \lambda; V\right),$$

so that i maps $B(V)$ to itself and in particular the fibers $F_{\text{aff}, \lambda}$, $\lambda \in \mathbb{C}$, to themselves. This is the analogue for two dimensions of the fact that one-dimensional Bloch varieties are hyperelliptic curves.

For all V in L^2, $B(V)$ is an affine hypersurface in $\mathbb{C}^* \times \mathbb{C}^* \times \mathbb{C}$. We construct a compactification $B(V)_{\text{comp}}$ of $B(V)$ such that π_{aff} extends to a morphism π from $B(V)_{\text{comp}}$ to the complex projective line P^1 and such that the generic points of $B(V)_{\text{comp}} \setminus B(V)$ are smooth points of $B(V)_{\text{comp}}$. Denote by $F_\lambda := \pi^{-1}(\lambda)$, $\lambda \in P^1$, the fibers of π. F_λ is the compactified Fermi curve over λ. Our main objective here is to compute the Fermi curve F_∞ over ∞.

THEOREM 1. $B(V)_{\text{comp}} \setminus B(V)$ is the union of the following curves

(1) Four curves $\Sigma^{e,f}$ $(e, f = 0, \infty)$ consisting of smooth points of $B(V)_{\text{comp}}$. They are sections of π and consist of smooth points of $B(V)_{\text{comp}}$.

(2) Four rational curves $Q^{e,f}$ $(e, f = 0, \infty)$, each of which has $(a-1)(b-1)$ ordinary double points. These curves are independent of V. $B(V)_{\text{comp}}$ is smooth at all points of $Q^{e,f} \setminus \text{Sing } Q^{ef}$.

(3) Four reduced divisors H_1^e, H_2^f $(e, f = 0, \infty)$. H_1^e (resp. H_2^f) is a hyperelliptic curve of arithmetic genus $a-1$ (resp. $b-1$). Specifically, H_1^e $(e = 0, \infty)$ is isomorphic to the one-dimensional Bloch variety.

$B(V_1) = \{(\xi, \lambda) \in \mathbb{C}^* \times \mathbb{C} |$ there is a complex valued

function ψ on Z satisfying

$$- \psi(m-1) - \psi(m+1) + V_1(m)\psi(m) = \lambda\psi(m)$$

and $\psi(m+a) = \xi\psi(m)$ for all m in $Z\}$,

where $V_1(m) = (1/b)\sum_{n=1}^b V(m, n)$. Similarly, H_2^f $(f = 0, \infty)$ is isomorphic to $B(V_2)$ where $V_2(n) = (1/a)\sum_{m=1}^a V(m, n)$.

The Fermi curve F_∞ over infinity is the transverse union of these eight curves:

$$F_\infty = \bigcup_{e, f = 0, \infty} Q^{ef} \cup \bigcup_{e=0, \infty} H_1^e \cup \bigcup_{f=0, \infty} H_2^f.$$

It is reduced. The components of F_∞ intersect only at smooth points of $B(V)_{\text{comp}}$. Their intersection pattern is indicated by the graph in Figure 1.

The compactification described above is constructed by embedding $\mathbb{C}^2 \times \mathbb{C}$ in projective space and successive blowups. In [**Bä**] there is a more natural construction by means of a toroidal embedding.

Theorem 1 (together with its proof) has a number of important consequences. For example:

Let $V \in L^2(Z^2/a\,Z \oplus b\,Z)$. Then

(1) $B(V)$ is irreducible.

(2) For each $\lambda \in \mathbb{C}$, $F_{\text{aff}, \lambda} = \pi_{\text{aff}}^{-1}(\lambda)$ is reduced. There are only finitely many points λ in $\mathbb{C}$ for which $F_{\text{aff}, \lambda}$ is reducible, and for these points $F_{\text{aff}, \lambda}$ consists of only two components.

(3) The arithmetic genus of a compactified Fermi curve F_λ is $2ab - 1$.

(4) If V is real, then for generic λ the Fermi curve F_λ has at most ordinary double points as singularities.

(5) *If V is real and the generic Fermi curve has $2ab - 2$ ordinary double points, that is, if its normalization is elliptic, then V is constant.*

Statement (5) is an analogue of a theorem of Borg [**Bo**] for the one-dimensional periodic Schrödinger equation. Borg showed that the spectrum of $-(d^2/dx^2) + q(x)$, q a real-valued function in $L^2(R/Z)$, as an operator on $L^2(R^1)$ is connected if and only if q is constant. (The same fact may be proven for the discrete one-dimensional periodic Schrödinger equation.) It is not hard to see that Borg's result may be reformulated in geometric terms as the statement that q is constant if and only if the normalization of the associated (one-dimensional) Bloch variety $B(q)$ is isomorphic to P^1. Now we may consider the minimal resolution of the surface $B(V)$, $V \in L^2(Z^2/aZ \oplus bZ)$. Statement (5) can easily be recast as the:

TWO-DIMENSIONAL BORG'S THEOREM. *Suppose V is real. Then the minimal resolution of $B(V)$ is a family of elliptic curves if and only if V is constant.*

There is another important aspect of Theorem 1. Let $\rho = (\rho_1, \rho_2) \in \mu_a \times \mu_b$ where μ_a (μ_b) denotes the multiplicative group of the ath (bth) roots of unity. The Fourier coefficient $\hat{V}(\rho)$ is

$$\hat{V}(\rho) := \frac{1}{ab} \sum_{\substack{1 \le m \le a \\ 1 \le n \le b}} V(m, n)\rho_1^{-m}\rho_2^{-n}.$$

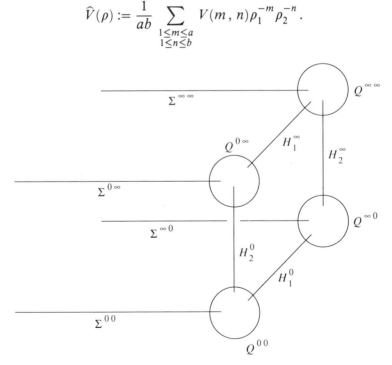

FIGURE 1

Of course

$$V(m, n) = \frac{1}{ab} \sum_{\rho \in \mu_a \times \mu_b} \widehat{V}(\rho) \rho_1^m \rho_2^n.$$

Each of the components $Q^{e,f}$ ($e, f = 0, \infty$) of F_∞ has $ab - (a + b) + 1$ ordinary double points. In the course of proving Theorem 1, one sees that each of these singularities is naturally labelled by a point $\rho = (\rho_1, \rho_2)$ in $\mu_a \times \mu_b$ with ρ_1 and ρ_2 different from 1. One can also see that there are coordinates x_1, x_2, x_3 (independent of V) in a neighborhood U_ρ of the point of $\operatorname{Sing} Q^{e,f}$ labelled by ρ such that $U_\rho \cap B(V)_{\text{comp}}$ is the zero-set of an equation of the form

$$ax_1^2 = x_2^2 + x_3^2 + \text{ terms of higher order},$$

where $a = \widehat{V}(\rho) \cdot \widehat{V}(\overline{\rho})$. Using the Nullstellensatz one deduces that for all potentials V the numbers $\widehat{V}(\rho) \cdot \widehat{V}(\overline{\rho})$ where $\rho = (\rho_1, \rho_2)$ with $\rho_1 \neq 1$, $\rho_2 \neq 1$ can be read off from the geometry of $B(V)$. Observe that for real-valued potentials $\widehat{V}(\overline{\rho}) = \overline{\widehat{V}(\rho)}$, so that in this case $B(V)$ determines the absolute values of all the Fourier coefficients $\widehat{V}(\rho)$, $\rho_1 \neq 1$, $\rho_2 \neq 1$. In part 3 of Theorem 1 above we have seen that $B(V)$ also determines the isospectral class of the averaged potentials V_1 and V_2.

Is there more information about the potential V contained in the geometry of $B(V)$? The best possible result would be that $B(V)$ determines V up to the obvious symmetries, namely translation and reflection in the lattice. The example of "separable potentials" $V(m, n) = V_1(m) + V_2(n)$ shows that this cannot be true in general, because in such a case $B(V)$ is determined by $B(V_1)$ and $B(V_2)$ (see §3), and the isospectral class of most one-dimensional potentials is nontrivial.

However, it is generically true that $B(V)$ determines V up to symmetries.

THEOREM 2. *There is a Zariski-open and dense subset P of L^2 such that for any $V \in P$, $V' \in L^2$ equality $B(V) = B(V')$ of the Bloch varieties implies that V and V' coincide up to symmetries, i.e., that there are $(m_0, n_0) \in Z/aZ \oplus bZ$ and such that $V'(m, n) = V(\pm m + m_0, \pm n + n_0)$.*

Recall that a Zariski-open subset of L^2 is the complement of the zero-set of one or several polynomial functions on L^2. Therefore the intersection of P with the set L_R^2 of real valued potentials is dense in L_R^2.

In contrast to most of the other results described in this announcement the proof of this theorem fundamentally relies on the algebra-geometric set-up. We consider the map from L^2 to the space of normalized polynomials in one variable of degree ab that associates to $V \in L^2$ the polynomial $P(1, 1, \lambda)$ whose zero-set is the periodic spectrum of $-\Delta + V$. Using compactifications of L^2 and the space of polynomials and a degeneration of $(-\Delta + V)$ to the multiplication operator V one sees that the generic fiber of this map is finite, and that the function $\hat{V}(\rho) \cdot \hat{V}(\overline{\rho})$ generically take different values on two elements V, $V' \in L^2$ that lie in one fiber of this map but are not related by a symmetry of the lattice.

3. Separable Bloch varieties

Suppose $V = V_1(m) + V_2(n)$, where $V_1 \in L^2(Z/aZ)$ and $V_2 \in L^2(Z/bZ)$. Then, the corresponding Bloch variety $B(V)$ is termed separable since the spectral problem

$$(-\Delta + V_1(m) + V_2(n))\psi = \lambda\psi,$$
$$\psi(m+a,n) = \xi_1\psi(m,n), \qquad \psi(m,n+b) = \xi_2\psi(m,n),$$

for all (m,n) in Z^2 can be solved by separation of variables. It is important to analyze these special cases in great detail because, as explained in the next section, we establish properties of the generic Bloch variety by perturbing about them. Besides, they are interesting in themselves.

It is convenient to introduce the unramified covering

$$c : \mathbb{C}^* \times \mathbb{C}^* \times \mathbb{C} \longrightarrow \mathbb{C}^* \times \mathbb{C}^* \times \mathbb{C}$$
$$(z_1, z_2, \lambda) \longrightarrow (z_1^a, z_2^b, \lambda),$$

and the preimage $\tilde{B}(V) := c^{-1}(B(V))$ of the Bloch variety $B(V)$ for any V in $L^2(Z^2/aZ \oplus bZ)$. Of course, $B(V)$ is the quotient of $\tilde{B}(V)$ by the action of $\mu_a \times \mu_b$ on $\mathbb{C}^* \times \mathbb{C}^* \times \mathbb{C}$ given by $\rho \cdot (z_1, z_2, \lambda) = (\rho_1 z_1, \rho_2 z_2, \lambda)$ for ρ in $\mu_a \times \mu_b$.

Let us begin our discussion of the separable case with the lifted Bloch variety $\tilde{B}(0)$ for $V = 0$. Set $e_\rho(m,n;z_1,z_2) = (\rho_1 z_1)^m (\rho_2 z_2)^n$, where $\rho \in \mu_a \times \mu_b$ and $(z_1, z_2) \in \mathbb{C}^* \times \mathbb{C}^*$. Then

$$\Delta e_\rho = \left(\rho_1 z_1 + \frac{1}{\rho_1 z_1} + \rho_2 z_2 + \frac{1}{\rho_2 z_2}\right).$$

Furthermore, e_ρ, $\rho \in \mu_a \times \mu_b$, is a basis for the vector space of all functions on Z^2 satisfying

$$\psi(m+a,n) = z_1^a \psi(m,n), \qquad \psi(m,n+b) = z_2^b \psi(m,n),$$

for all (m,n) in Z^2. Thus

$$\tilde{B}(0) = \left\{(z_1, z_2, \lambda) \in \mathbb{C}^* \times \mathbb{C}^* \times \mathbb{C} \,\middle|\, \right.$$

$$\left. \prod_{\rho \in \mu_a \times \mu_b} \left(\rho_1 z_1 + \frac{1}{\rho_1 z_1} + \rho_2 z_2 + \frac{1}{\rho_2 z_2} + \lambda\right) = 0\right\}.$$

Let

$$E := \left\{(z_1, z_2, \lambda) \in \mathbb{C}^* \times \mathbb{C}^* \times \mathbb{C} \,\middle|\, z_1 + \frac{1}{z_1} + z_2 + \frac{1}{z_2} + \lambda = 0\right\},$$

and

$$\rho \cdot E := \left\{ (z_1, z_2, \lambda) \in \mathbb{C}^* \times \mathbb{C}^* \times \mathbb{C} \, \middle| \, \frac{z_1}{\rho_1} + \frac{\rho_1}{z_1} + \frac{z_2}{\rho_2} + \frac{\rho_2}{z_2} + \lambda = 0 \right\},$$

for ρ in $\mu_a \times \mu_b$. Then

$$\widetilde{B}(0) = \bigcup_{\rho \in \mu_a \times \mu_b} \rho \cdot E.$$

Finally, denote by E_λ the fiber over λ of the projection of E on the λ-plane.

The closure of E_λ, $\lambda \neq 0, \pm 4$, in $P^1 \times P^1$ is an elliptic curve. If $\lambda = 0$, E_λ splits into two components $z_1 = -z_2$ and $z_1 = -\frac{1}{z_1}$. If $\lambda = \pm 4$, E_λ is irreducible with $\pm(1, 1)$ as its only singular point. The closure of E in $P^1 \times P^1 \times P^1$ with its projection onto the third factor is, after blowing up four points lying over $\lambda = \infty$, a stable family of elliptic curves over P^1 with four exceptional fibers; in Kodaira's notation, two of type I_1, one of type I_2 and one of type I_7. So, by [**Be**] it is a modular family and its monodromy group is isomorphic to $\Gamma_0(8) \cap \Gamma_0^0(4)$.

We see that $\widetilde{B}(0)$ is the union of $a \cdot b$ components each of which is the translate of the family of elliptic curves E by an element of $\mu_a \times \mu_b$. In particular, the lifted Fermi curve $\widetilde{F}_\lambda$ for $V = 0$ is

$$\widetilde{F}_\lambda = \bigcup_{\rho \in \mu_a \times \mu_b} \rho \cdot E_\lambda.$$

To analyze the topology of a generic family of Fermi curves via degeneration to $V = 0$ requires considerable information about the singularities of $B(0)$ and its fibers F_λ. Since $B(0) = \widetilde{B}(0)/\mu_a \times \mu_b$ is a quotient, the singularity structure may be expressed in terms of intersections of translates of E. Some of this information is contained in

THEOREM 1. *For a and b large enough distinct odd primes,*

(1) *Any two distinct translates of E are transversal.*

(2) *Any three distinct translates of E are in general position at every point of intersection, i.e., the normals to their tangent planes are linearly independent.*

(3) *The possible singularities of F_λ are:*

    (i) *ordinary double points (Figure 2),*

FIGURE 2

(ii) *ordinary triple points* (*Figure* 3),

FIGURE 3

(iii) *ordinary quadruple points* (*Figure* 4),

FIGURE 4

(iv) *ordinary tacnode points* (*Figure* 5),

FIGURE 5

(v) *"spectral" quadruple points* (*Figure* 6),

FIGURE 6

Configuration (v) *only occurs at points* (ξ_1, ξ_2, λ) *on* $B(0)$ *with* $\xi_1^2 = \xi_2^2 = 1$.

(4) *Five distinct translates of* E *never intersect. In other words there are no quintuple points.*

Theorem 1 is not enough to successfully control the perturbation $B(\epsilon V)$ about $B(0)$ for generic V in $L^2(Z^2/aZ \oplus bZ)$. One needs to know for example, that ordinary triple and quadruple points are "independent with respect to deformation", as for example they would be if they did not occur in the same fiber. Unfortunately, the precise meaning we give to "independent with respect to deformation" is rather technical, so we don't include it here. This completes our synopsis of $V = 0$.

We now discuss the structure of $B(V)$ for $V = V_1 + V_2$. Let $B(V_i) \subset \mathbb{C}^* \times \mathbb{C}$, $i = 1, 2$, be the one-dimensional Bloch varieties for V_j, $j = 1, 2$. Suppose $(\xi_i, \lambda_i) \in B(V_i)$, $i = 1, 2$, and ψ_i, $i = 1, 2$, are the corresponding eigenfunctions. Then, $\psi = \psi_1(m)\psi_2(n)$ is an eigenfunction of $-\Delta + V$ with eigenvalue $\lambda_1 + \lambda_2$ satisfying $\psi(m+a, n) = \xi_1 \psi(m, n)$ and $\psi(m, n+b) = \xi_2 \psi(m, n)$. Therefore, the image of the map

$$B(V_1) \times B(V_2) \xrightarrow{j} \mathbb{C}^* \times \mathbb{C}^* \times \mathbb{C}$$
$$((\xi_1, \lambda_1), (\xi_2, \lambda_2)) \longrightarrow (\xi_1, \xi_2, \lambda_1 + \lambda_2),$$

is $B(V)$. The next theorem describes the singular locus of $B(V_1 + V_2)_{comp}$ for generic V_1 and V_2.

THEOREM 2. *Let* $V_1 \in L^2(Z/aZ)$ *and* $V_2 \in L^2(Z/bZ)$ *be generic one dimensional potentials. The Bloch variety* $B = B(V_1 + V_2)_{comp}$ *has the following properties*:

(1) *B is mapped to itself by the two involutions* $i_1(\xi_1, \xi_2, \lambda) = (1/\xi_1, \xi_2, \lambda)$ *and* $i_2(\xi_1, \xi_2, \lambda) = (\xi_1, 1/\xi_2, \lambda)$. *Of course* $i = i_1 \circ i_2$ *and the group I generated by i_1 and i_2 is isomorphic to* $Z/2Z \times Z/2Z$.

(2) *The singular locus* $\text{Sing}(B)$ *is an ordinary double curve on B. It is irreducible. The restriction of π to* $\text{Sing}(B)$ *is a branched covering of P^1 of degree* $2(a-1)(b-1)$. *If λ_0 is a regular value of* $\pi|_{\text{Sing}(B)}$ *then for any four points* p_1, p'_1, p_2, p'_2 *in* $\text{Sing}(B) \cap F_{\lambda_0}$ *with* $p'_i \notin I \cdot p_j$, $i, j = 1, 2$, *there is an element w in the covering group of* $\pi|_{\text{Sing}(B)}$ *such that* $w(p_1) = p_2$ *and* $w(p'_1) = p'_2$.

(3) *There are finite subsets* $G_0 \subset B(V_1 + V_2)_{aff} \setminus \text{Sing}(B)$, $G_1, G_2, G_3, G_4 \subset B(V_1 + B_2)_{aff} \cap \text{Sing}(B)$ *such that*

   (i) *The sets* $D_i = \pi(G_i)$, $i = 0, \ldots, 4$, *are disjoint. The projection map π induces a bijection between G_i/I and D_i.*

   (ii) *G_0 is the fixed point set of the involution i on B and consists of $4ab$ different points. For each p in G_0 the Fermi curve* $F_{\pi(p)}$ *has an ordinary double point at p.*

   (iii) *For λ in* C, $\text{Sing}(F_\lambda) = (F_\lambda \cap \text{Sing}(B)) \cup (G_0 \cap F_\lambda)$. *If $p \in \text{Sing}(B) \setminus (G_1 \cup G_2 \cup G_3 \cup G_4)$, then p is an ordinary double point of* $F_{\pi(p)}$.

   (iv) *For each $p \in G_1$ one can introduce coordinates* $x = x(\xi_1, \xi_2, \lambda)$, $y = y(\xi_1, \xi_2, \lambda)$, $\mu = \mu(\lambda)$ *around p in* $C^* \times C^* \times C$ *such that in these coordinates B is given by the equation* $y^2 = x^3 + \mu x^2$. *See Figure 7.*

$$\mu = 0 \qquad\qquad\qquad\qquad \mu \neq 0$$

FIGURE 7

I acts without fixed points on G_1 and D_1 consists of
$(a - 1)(b - 1)$ *points.*

   (v) *For each p in G_2 one can introduce coordinates x, y and μ around p in* $C^* \times C^* \times C$ *such that B is given by* $(x + y^2) \times (x - y^2 - \mu) = 0$. *See Figure 8.*

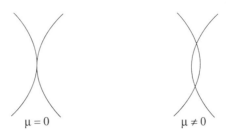

FIGURE 8

I acts without fixed points on G_2 and D_2 has
$(1/2)(a - 1)(b - 1)(2a - 2b - 7)$ elements.

FIGURE 9

I acts without fixed points on G_3 and D_3 consists of
$(1/6)(a - 1)(a - 2)(b - 1)(b - 2)$ points.

(vi) *For each p in G_3 one can introduce coordinates x, y and μ*
around p such that B is given by $xy(x+y+\mu) = 0$. See Figure 9.

(vii) *For each p in G_4 there are coordinates x, y and μ around*
p such that B is given by $y^2 = x^3 + \mu x^2$. The isotropy group
of a point in G_4 is either $\langle i_1 \rangle$ or $\langle i_2 \rangle$. D_4 consists of $(a - 1)$
$\times (b - 1)(a + b)$ points.

Theorem 2 shows that the generic Fermi curve of $B(V_1 + V_2)$ has $2(a - 1)(b - 1)$ double points. Parts (iv) through (vii) of (3) describe the nature of the singularities on the Fermi curves which are not ordinary double points. There are ordinary cusps (iv), tacnodes (v) and (vii), and ordinary triple points (vi). Thus, the singularities of $B(V_1 + V_2)$ are generally simpler than those of $B(0)$. The theorem also provides normal forms for the deformation of these singularities represented by the family of Fermi curves.

4. Topology of the generic Bloch variety

For an arbitrary potential $V \in L^2(Z^2/a Z \oplus b Z)$ we consider the family of curves $\pi : B(V)_{\text{comp}} \to P^1$. By the Ehresmann fibration theorem there is a finite subset $D \subset P^1$ such that π is a locally trivial fiber bundle over $P_1 \backslash D$. The points of D are called *van Hove singularities*. If $B(V)$ is smooth then (by the theorem of Lefschetz-Sard) almost all fibers of π are smooth.

In this case the set of van have singularities is the set of points $\lambda \in P^1$ for which the Fermi curve F_λ of π is singular. This is the typical situation, and for a generic potential V the family $\pi : B(V)_{\text{comp}} - F_\infty \to P^1$ is almost a Lefschetz pencil.

THEOREM 1. *There is a dense (Zariski-) open subset $\mathscr{E}$ of $L^2(Z^2/aZ \oplus bZ)$ such that for all $V \in \mathscr{E}$ the affine Bloch variety $B(V)$ is smooth and the set D of van Hove singularities consists of $\nu_{a,b} = 2a^2b^2 - 6(a^2 + b^2) + 36(a + b) + 1$ points. In this case the singularities of a Fermi curve F_λ, $\lambda \in D \setminus \{\infty\}$ are either one or two ordinary double points. If F_λ has two ordinary double points then they are interchanged by the involution i.*

If $\lambda \neq \infty$ is a van Hove singularity whose Fermi curve contains only one singular point then this point is left fixed by the involution i, so it is of the form (ξ_1, ξ_2, λ) with $\xi_1, \xi_2 \in \{\pm 1\}$. We then call λ a *spectral van Hove singularity* since it belongs to the spectrum of $-\Delta + V$ with periodic, antiperiodic or mixed periodic-antiperiodic boundary conditions. It is also easy to see that any such point in the spectrum is a spectral van Hove singularity; i.e., every point (ξ_1, ξ_2, λ) of $B(V)$ with $\xi_1, \xi_2 \in \{\pm 1\}$ is a singular point of π.

We also point out that in the situation of Theorem 2 in §3 all singular points of π on $B(V)$ are ordinary Morse singularities. In other words one can introduce coordinates $x_1 = x_1(\xi_1, \xi_2, \lambda)$, $x_2 = x_2(\xi_1, \xi_2, \lambda)$, $\mu = \mu(\lambda)$ in a neighborhood of such a point so that in this neighborhood $B(V)$ is the zero set of the equation $\mu = x_1^2 + x_2^2$.

This normal form for the singularities of π on $B(V)$, $V \in \mathscr{E}$ is a direct consequence of the fact that a one-parameter deformation of an ordinary double point has a smooth total space if and only if it is of the form above.

Though Theorem 1 is not unexpected its proof is technically quite complicated. Roughly, the proof consists in first showing that for a generic potential the Bloch variety has all the topological properties mentioned in the theorem. Then, computing various Euler characteristics, one sees that any such potential has the desired number ν_{ab} of van Hove singularities. Conversely, for any potential V with a smooth Bloch variety and ν_{ab} van Hove singularities $B(V)$ can be connected to the Bloch variety of a generic potential in a topologically trivial family, and the theorem follows.

In order to show that for a generic potential V the Bloch variety has all the topological properties of Theorem 1 it suffices to exhibit one potential for which this is the case, because the failure of these properties is a Zariski-closed condition. Unfortunately, we are not able to explicitly write down such a potential. Therefore we prove its existence by perturbation theory around the zero-potential. In other words, for $V \in L^2$ we consider the family $B(\epsilon V)$, ϵ in $\mathbb{C}$ of Bloch varieties, and show that for generic V and sufficiently small $\epsilon \neq 0$ the Bloch variety $B(\epsilon V)$ has all the required properties.

There are two aspects to this approach. One is to show that the Bloch variety is smooth and the Fermi curves have only ordinary double points as singularities, and the other is to show that there are at most two singularities on a Fermi curve F_λ, $\lambda \neq \infty$. The second aspect requires some global arguments, in particular the fact that the singularities of $B(0)$ described in (ii)–(v) of part 3 of Theorem 1 of §3 are "independent with respect to deformation". We do not discuss this second aspect here and only give some indication how the first problem is overcome. Using the compactification of §2 we may reduce everything to showing that any point p of $B(0)$ has a neighborhood U in $\mathbb{C}^2 \times \mathbb{C}$ such that for generic $V \in L^2$ and small ϵ in $\mathbb{C} \setminus \{0\}$ the intersection $B(\epsilon V) \cap U$ is smooth and that $\pi|_{B(\epsilon V) \cap U}$ has only Morse-type singularities.

To obtain local equations for $B(\epsilon V)$ it is convenient to use the unramified cover $c : \mathbb{C}^* \times \mathbb{C}^* \times \mathbb{C} \to \mathbb{C}^* \times \mathbb{C}^* \times \mathbb{C}$, $(z_1, z_2, \lambda) \to (z_1^a, z_2^b, \lambda)$ introduced in §3. We have seen that for any $(z_1, z_2) \in \mathbb{C}^* \times \mathbb{C}^*$ the functions $e_\rho(z)$: $(m, n) \to (\rho_1 z_1)^m (\rho_2 z_2)^n$, $\rho \in \mu_a \times \mu_b$ form a basis of the space of functions ψ on Z^2 satisfying $\psi(m+a, n) = z_1^a \psi(m, n)$, $\psi(m, n+b) = z_2^b \psi(m, n)$. We have also seen that with respect to this basis the operator $-\Delta - \lambda$ is represented by the diagonal $ab \times ba$-matrix (indexed by $\mu_a \times \mu_b$)

$$[-\Delta - \lambda]_{\rho, \rho'} = -\left(\rho_1 z_1 + \frac{1}{\rho_2 z_2} + \rho_2 z_2 + \frac{1}{\rho_2 z_2} + \lambda\right)\delta_{\rho, \rho'}.$$

It is easy to see that the multiplication operator V is represented in this basis by the "convolution matrix":

$$[V]_{\rho, \rho'} = \widehat{V}(\rho \cdot \overline{\rho}').$$

Clearly

$$\det([-\Delta - \lambda]_{\rho, \rho'} + [V]_{\rho, \rho'}) =: \widetilde{P}_V(z_1, z_2, \lambda),$$

is an equation for $\widetilde{B}(V)$.

Now fix a point $\tilde{p} \in \widetilde{B}(0)$, and put $\lambda_0 := \tilde{\pi}(p)$, where $\tilde{\pi} := \pi \circ c$. We want to show that there is a neighborhood $\widetilde{U}$ of $\tilde{p}$ such that for generic $V \in L^2$ and sufficiently small $\epsilon \neq 0$ the intersection $\widetilde{U} \cap \widetilde{B}(\epsilon V)$ is smooth and that the restriction of $\tilde{\pi}$ to this set has only Morse type singularities. This is clearly the case if $\tilde{p}$ itself is a smooth point of $\widetilde{B}(0)$ and a regular point for $\tilde{\pi}|_{\widetilde{B}(0)}$. In all other cases we make a deformation argument that depends on the nature of the singularity of $\bigcup_{\rho \in \mu_a \times \mu_b} E_{\lambda_0}$ at $\tilde{p}$. Remember that all possible types of singularities that can occur have been enumerated in part 3 of Theorem 1 in §3. We sketch the deformation argument in two of these cases.

First consider the case that $\tilde{p}$ lies on precisely two components of $\widetilde{B}(0)$, and that the restriction of $\tilde{\pi}$ to any of these components has $\tilde{p}$ as a regular point. In other words we assume that $\tilde{p}$ is a general point of a double curve on $\widetilde{B}(0)$. Using a covering transformation we may assume that $\tilde{p}$ lies on E

and one translate, say $\sigma \cdot E$, of E. Let us order $\mu_a \times \mu_b$ in such a way that $1 = (1, 1)$ and σ are the first two elements. If the average $\widehat{V}(0)$ of V is equal to zero then the matrix $[-\Delta - \lambda]_{\rho, \rho'} + [\epsilon V]_{\rho, \rho'}$ has the form

$$\begin{array}{c} \longrightarrow \rho \\ \rho' \downarrow \end{array} \left(\begin{array}{c|c} A & B \\ \hline C & D \end{array} \right) ,$$

$$A = \begin{array}{c} \\ 1 \\ \sigma \end{array} \begin{array}{cc} \overset{1}{} & \overset{\sigma}{} \end{array} \left(\begin{array}{cc} -\left(z_1 + \frac{1}{z_1} + z_2 + \frac{1}{z_2} + \lambda \right) & \epsilon \widehat{V}(\sigma) \\ \epsilon \widehat{V}(\overline{\sigma}) & -\left(\sigma_1 z_1 + \frac{1}{\sigma_1 z_1} + \sigma_2 z_2 + \frac{1}{\sigma_2 z_2} + \lambda \right) \end{array} \right) ,$$

$$B = \overset{\longrightarrow \rho}{\left(\begin{array}{ccccccc} \cdot & \cdot & \cdot & \epsilon \widehat{V}(\rho) & \cdot & \cdot & \cdot \\ \cdot & \cdot & \cdot & \epsilon \widehat{V}(\rho\overline{\sigma}) & \cdot & \cdot & \cdot \end{array} \right)} ,$$

$$C = \rho' \downarrow \left(\begin{array}{cc} \vdots & \vdots \\ \epsilon \widehat{V}(\overline{\rho}') & \epsilon \widehat{V}(\overline{\sigma}\rho') \\ \vdots & \vdots \end{array} \right) ,$$

$$D = \left(\begin{array}{ccc} & & \epsilon V(\rho\overline{\rho}') \\ & -\left(p_1 z_1 + \frac{1}{p_1 z_1} + p_2 z_2 + \frac{1}{p_2 z_2} + \lambda \right) & \\ \epsilon \widehat{V}(\rho'\overline{\rho}) & & \end{array} \right) .$$

By the assumptions on $\tilde{p}$

$$x_1 := -\left(z_1 + \frac{1}{z_1} + z_2 + \frac{1}{z_2} + \lambda \right) ,$$

$$x_2 := -\left(\sigma_1 z_1 + \frac{1}{\sigma_1 z_1} + \sigma_2 z_2 + \frac{1}{\sigma_2 z_2} + \lambda \right) ,$$

$$\mu := \lambda - \lambda_0 ,$$

are coordinates in a neighborhood U of $\tilde{p}$. After shrinking U we may assume that for $\rho \neq 1$ or σ the function

$$f_\rho(x_1, x_2, \mu) := -\left(p_1 z_1 + \frac{1}{p_1 z_1} + p_2 z_2 + \frac{1}{p_2 z_2} + \lambda \right) ,$$

does not vanish anywhere on $\widetilde{U}$. Then

$$\left(\prod_{\substack{\rho \in \mu_a \times \mu_b \\ \rho \neq \sigma, 1}} f_\rho^{-1} \right) \cdot \widetilde{P}_{\epsilon V} ,$$

is an equation for $\tilde{B}(\epsilon V) \cap \tilde{U}$. Expanding the determinant of the matrix above one sees that this equation has the form

$$(*) \qquad\qquad x_1 x_2 - \epsilon^2 \hat{V}(\sigma)\hat{V}(\overline{\sigma}) + \epsilon^2 h(x, \mu, \epsilon),$$

where h is a function with $h(\tilde{p}, 0) = 0$. Whenever $\hat{V}(\sigma) \cdot \hat{V}(\overline{\sigma}) \neq 0$ it follows by elementary estimates that the zero-set of $(*)$ in $\tilde{U}$ is smooth for all small $\epsilon \neq 0$, and that the map $(x_1, x_2, \mu) \to \mu$ has no critical points on this zero set. This gives the desired result and also shows that for real potentials the absolute value of the Fourier-coefficient $\hat{V}(\sigma)$ governs the "rate of opening" of the double curve $E \cap \sigma \cdot E$. Similar calculations can be found in the physics literature, e.g., [AM, Chapter 9].

We now consider the case of a triple point (i.e., case 3 (ii) of Theorem 1 in §3). As before it may be assumed that there are $\sigma, \tau \in \mu_a \times \mu_b \setminus \{1\}$ with $\sigma \neq \tau$ such that $\tilde{p} \in E \cap \sigma E \cap \tau T$ and $\tilde{p} \notin \rho \cdot E$ for $\rho \neq \sigma, \tau, 1$. Again one can show that for generic V the geometry of $\tilde{B}(\epsilon V)$ is a neighborhood of $\tilde{p}$ is already accurately modelled by the zero-set of the determinant of the 3×3-subblock of the matrix $[-\Delta - \lambda]_{\rho, \rho'}$, associated to the indices $1, \sigma, \tau$:

$$\begin{pmatrix} -\left(z_1 + \frac{1}{z_1} + z_2 + \frac{1}{z_2} + \lambda\right) & \epsilon\hat{V}(\sigma) & \epsilon\hat{V}(\tau) \\ \epsilon\hat{V}(\overline{\sigma}) & -\left(\sigma_1 z_1 + \frac{1}{\sigma_1 z_1} + \sigma_2 z_2 + \frac{1}{\sigma_2 z_2} + \lambda\right) & \epsilon\hat{V}(\overline{\sigma}\tau) \\ \epsilon\hat{V}(\overline{\tau}) & \epsilon\hat{V}(\sigma\overline{\tau}) & -\left(\tau_1 z_1 + \frac{1}{\tau_1 z_1} + \tau_2 z_2 + \frac{1}{\tau_2 z_2} + \lambda\right) \end{pmatrix}.$$

One can also show that in a neighborhood U of $\tilde{p}$ one can introduce coordinates $x_1 = x_1(z_1, z_2, \lambda)$, $x_2 = x_2(z_1, z_2, \lambda)$, $u = \mu(\lambda)$ such that

$$x_1 = -\left(z_1 + \frac{1}{z_1} + z_2 + \frac{1}{z_2} + \lambda\right),$$

$$x_2 = -\left(\sigma_1 z_1 + \frac{1}{\sigma_1 z_1} + \sigma_2 z_2 + \frac{1}{\sigma_2 z_2} + \lambda\right),$$

$$x_1 + x_2 - \mu = -\left(\tau_1 z_1 + \frac{1}{\tau_1 z_1} + \tau_2 z_2 + \frac{1}{\tau_2 z_2} + \lambda\right).$$

Thus the problem is reduced to showing that there is a neighborhood U' of 0 in $\mathbb{C}^3$ such that for generic V and all sufficiently small $\epsilon \neq 0$ the zero-set of

$$\det \begin{pmatrix} x_1 & \epsilon\hat{V}(\sigma) & \epsilon\hat{V}(\tau) \\ \epsilon\tilde{V}(\overline{\sigma}) & x_2 & \epsilon\tilde{V}(\tau\overline{\sigma}) \\ \epsilon\hat{V}(\overline{\tau}) & \epsilon\hat{V}(\overline{\tau}\sigma) & x_1 + x_2 - \mu \end{pmatrix},$$

in U' is smooth and that the map $(x_1, x_2, \mu) \to \mu$ has only Morse singularities on this set. Though the Fourier coefficients appearing in this matrix are not always independent (for example one could have $\sigma = \tau\overline{\sigma}$) it is possible to show that there are always enough degrees of freedom to achieve the desired result.

5. Monodromy for generic potentials

For any potential $V \in L^2(\mathbb{Z}^2/a\mathbb{Z} \oplus b\mathbb{Z})$ we denote by $Y = Y(V)$ the quotient of $B(V)_{\text{comp}}$ by the involution i.

$$Y(V) = B(V)_{\text{comp}}/\langle i \rangle.$$

The projection $\pi : B(V)_{\text{comp}} \to P^1$ induces a morphism from $Y(V)$ to P^1 which (by abuse of notation) we also call π. The fibers $Y_\lambda = Y(V)_\lambda := \pi^{-1}(\lambda)$ are the quotients of the Fermi curves F_λ by i, $Y(V)_\lambda = F_\lambda/\langle i \rangle$.

From now on we assume that V lies in the set $\mathscr{E}$ of generic potentials described in §4. Then Y_λ is smooth for all λ outside the set D of van Hove singularities. If $\lambda \in D \setminus \{\infty\}$ then, by the theorem of §4, the curve Y_λ has precisely one singular point, and this singular point is an ordinary double point. One easily verifies that this double point of Y_λ is a smooth point of Y if λ is not a spectral van Hove singularity, and that it is an ordinary double point of the surface Y if λ is a spectral van Hove singularity.

Since $\pi : Y \to P^1$ is a locally trivial fiber bundle over $P^1 \setminus D$ every path w in $P^1 \setminus D$ lifts to a diffeomorphism $R(w) : Y_{\lambda_1} \to Y_{\lambda_2}$ between the fibers over the initial point λ_1 and the endpoint λ_2 of w. This diffeomorphism is only unique up to isotopy, but its isotopy class only depends on the homotopy class $[w]$ of w. Therefore, its induced map on homotopy

$$r([w]) : H_1(Y_{\lambda_1}, Z) \to H_1(Y_{\lambda_2}, Z),$$

is well defined. If we consider, in particular, paths starting and ending at a point $\lambda_0 \in P^1 \setminus D$ we get an action of the fundamental group $\pi_1(P^1 \setminus D, \lambda_0)$ of $P^1 \setminus D$ on $H_1(Y_{\lambda_0}, Z)$. It is called the *monodromy representation*

$$r : \pi_1(P^1 \setminus D, \lambda_0) \to \text{Aut}(H_1(Y_{\lambda_0}, Z)).$$

Its image is called the *monodromy group*.

The standard tool for computing the monodromy of families of algebraic varieties is the Picard-Lefschetz formula. In our context this amounts to the following. For $\lambda \in P^1 \setminus D$ the curve Y_λ is smooth, so the intersection form $\langle \, , \, \rangle$ on $H_1(Y_\lambda, Z)$ is a nondegenerate skew-symmetric bilinear form of determinant one. If $\mu \in D \setminus \{\infty\}$ is a van Hove singularity and $p \in Y_\mu$ is the unique singular point of Y_μ there is a diffeomorphism h between a neighborhood U of p in Y and a neighborhood U' of 0 in

$$\{((x_1, x_2, x_3) \in \mathbb{C}^3 | x_3 = x_1^2 + x_2^2\}, \quad \text{if } \mu \text{ is not a spectral}$$
$$\text{van Hove singularity, and}$$

$$\{((x_1, x_2, x_3) \in \mathbb{C}^3 | x_3^2 = x_1^2 + x_2^2\}, \quad \text{if } \mu \text{ is a spectral}$$
$$\text{van Hove singularity},$$

such that $\pi \circ h^{-1}(x) = f(x_3)$ with a function f such that $f(0) = \mu$, $f'(0) \neq 0$. See Figure 10.

For λ in a sufficiently small ball $K(\mu)$ around μ the set $\pi^{-1}(\lambda) \cap U$ has the homotopy type of a circle, so it represents a homology class $v(\mu)$

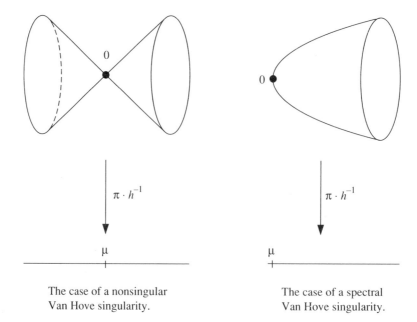

The case of a nonsingular   The case of a spectral
Van Hove singularity.   Van Hove singularity.

FIGURE 10

in $H_1(F_\lambda, Z)$. This homology class is unique up to sign. Its representative shrinks to p as $\lambda \to \mu$, so it is called the *vanishing cycle* for p (or also for μ). If w is a simple loop in $K(\mu)$ around μ starting and ending at λ then the *Picard-Lefschetz formulas* give for all $\gamma \in H_1(Y_\lambda, Z)$:

$$r([w])(\gamma) = \gamma - \langle v(\mu), \gamma \rangle \cdot v(\mu) \quad \text{if } \mu \text{ is a non-spectral}$$
$$\text{van Hove singularity},$$

$$r([w])(\gamma) = \gamma - 2\langle v(\mu), \gamma \rangle \cdot v(\mu) \quad \text{if } \mu \text{ is a spectral}$$
$$\text{vanHove singularity}.$$

We now choose base point λ_0 near to ∞ on the negative real axis. By Theorem 1.1 the fiber Y_∞ consists of four components—two hyperelliptic curves and two rational curves. Each of the four points $p_1, \ldots, p_4$ where a rational curve and a hyperelliptic curve meet is an ordinary double point of Y_∞ and a smooth point of Y. As above one defines the vanishing cycles to these points, and easily verifies that they are homologous. The resulting homology class in $H_1(Y_{\lambda_0}, Z)$ is denoted by γ_∞; again it is unique up to sign.

THEOREM 1. *Let V be a potential in $\mathscr{E}$, λ_0 a point on the negative real axis close to ∞. Then*

(i) *The monodromy representation $r : \pi_1(P^1 \setminus D, \lambda_0) \to \text{Aut}(H_1(Y_{\lambda_0}, Z))$ is absolutely irreducible, that is, the induced representation on $H_1(Y_{\lambda_0}, \mathbb{C})$ has no nontrivial invariant subspaces,*

(ii) *Let L be the smallest sublattice of $H_1(Y_{\lambda_0}, Q)$ that is left fixed by* *the action of* $\pi_1(P^1 \setminus D, \lambda_0)$ *and contains all the elements*

$$\frac{1}{2}(\gamma_\infty - r([w]) \cdot \gamma_\infty), \qquad w \in \pi_1(P^1 \setminus D, \lambda_0).$$

Then $L = H_1(Y_{\lambda_0}, Z)$.

This theorem is formulated for a base point λ_0 on the negative real axis close to ∞. Of course one could pass to any other base point $\lambda_0' \in P^1 \setminus D$. One would then choose a path w connecting λ_0' to λ_0 in $P^1 \setminus D$. Then the monodromy representations of $\pi_1(P^1 \setminus D, \lambda_0')$ and $\pi_1(P^1 \setminus D, \lambda_0)$ only differ by conjugation with $r([w])$. The first statement of the theorem would not change at all, while in the second statement the definition of γ_∞ would have to be adjusted.

The proof of the theorem is by degeneration to a generic separable potential. Let us sketch it. Consider a one parameter family $V_s + \epsilon V_g$, ϵ in C, of potentials where V_s is a generic separable potential as in Theorem 2 in §3, and V_g is chosen such that $V_s + \epsilon V_g \epsilon \mathscr{E}$ for all but finitely many ϵ in C. Once the theorem is proven for $V = V_s + \epsilon V_g$ where ϵ is small it is clear that it holds for all potentials in the connected set $\mathscr{E}$ ($\mathscr{E}$ is connected since it is Zariski-open).

By Theorem 2 in §3 the curve $Y(V_s)_{\lambda_0}$ has precisely $(a-1)(b-1)$ ordinary double points. In the same way as above one associates to each of these double points a homology class in $H_1(Y(V_s + \epsilon V_g)_{\lambda_0}, Z)$ that vanishes at this point as $\epsilon \to 0$. We call these classes the $(\epsilon = 0)$-vanishing cycles in $H_1(Y(V_s + \epsilon V_g)_{\lambda_0}, Z)$. Observe that for any two $(\epsilon = 0)$-vanishing cycles v_1, v_2 there is $[w]$ in $\pi_1(P^1 \setminus D, \lambda_0)$ such that $v([w])(v_1) = v_2$. For w one can choose a path in $P^1 \setminus D$ such that the covering transformation of $\pi|_{\text{Sing } Y(V_s)} : \text{Sing } Y(V_s) \to P^1$ associated to w maps the point at which v_1 vanishes to the point at which v_2 vanishes. Such a path exists since $\text{Sing } Y(V_s)$ is irreducible (Theorem 2 in §3 part (3)(ii)).

Let F^0 be the subspace of $H_1(Y(V)_{\lambda_0}, Z)$ generated by the $(\epsilon = 0)$-vanishing cycles. Then the quotient $H_1(Y(V)_{\lambda_0}, Z)/F^0$ is canonically isomorphic to $H_1(Y(V_s)_{\lambda_0}, Z)$. The monodromy of the family $\pi : Y(V_s) \to P^1$ can be computed quite explicitly. The information about the monodromy of this family and about the $(\epsilon = 0)$-vanishing cycles is put together by studying the two-parameter-family $Y(V_2 + \epsilon V_g)_\lambda$ indexed by λ and ϵ around a cusp point of a fiber of $\pi : Y(V_s) \to P^1$ (Theorem 2 in §3 part (3)(iv)). Some of the vanishing cycles occurring in this deformation are $(\epsilon = 0)$-vanishing cycles and others are related to the homology of the separable family. The Picard-Lefschetz formulas can be used to see that they are related under the action of the monodromy group.

6. Density of states

Let V in $L^2(Z^2/aZ \oplus bZ)$ be *real*. Then, for each $k \in R^2$,

$$(-\Delta + V)\psi = \lambda\psi,$$

$$\psi(m+a, n) = e^{ik_1}\psi(m, n), \qquad \psi(m, n+b) = e^{ik_2}\psi(m, n),$$

for all $(m, n) \in Z^2$, is a self adjoint boundary value problem. Denote its spectrum as before by

$$E_1(k) \le E_2(k) \le \cdots \le E_{ab}(k).$$

The function $k \to E_j(k)$, $1 \le j \le ab$, on R^2 is again called the jth band function. It is piecewise real-analytic and periodic with respect to $2\pi Z^2$. For each $1 \le j \le ab$ the "graph" of E_j

$$\{(e^{ik_1}, e^{ik_2}, E_j(k))|k \in R^2\},$$

is contained in $B(V)$. It follows at once from the fact that $B(V)$ is irreducible that the analytic continuation of any single band function determines $B(V)$ and in particular all other band functions.

Let H_n, $n \ge 1$, denote the selfadjoint operator $-\Delta + V$ acting on $L^2(Z^2/anZ \oplus bnZ)$. For $k \in \frac{2\pi}{n}Z \oplus \frac{2\pi}{n}Z$, $E_j(k)$, $1 \le j \le ab$, is an eigenvalue of H_n. It follows that the spectrum of H_n is

$$\left\{ E_j\left(\frac{2\pi}{n}m_1, \frac{2\pi}{n}m_2\right) \middle| 1 \le j \le ab, \ 1 \le m_1, \ m_2 \le n \right\}.$$

Let $\nu_n(\lambda)$, $n \ge 1$, be the number of eigenvalues of H_n less than or equal to λ. We may write

$$\nu_n(\lambda) = \sum_{j=1}^{a \cdot b} \sum_{m_1, m_2=1}^{n} \Theta\left(\lambda - E_j\left(\frac{2\pi m_1}{n}, \frac{2\pi m_2}{n}\right)\right),$$

where

$$\Theta(t) = \begin{cases} 1, & t \ge 0 \\ 0, & t < 0, \end{cases}$$

is the Heaveyside function. Therefore, the integrated density of states

$$\rho(\lambda) := \lim_{n \to \infty} \frac{1}{abn^2}\nu_n(\lambda)$$

$$= \lim_{n \to \infty} \frac{1}{ab}\sum_{j=1}^{a \cdot b} \frac{1}{n^2} \sum_{m_1, m_2=1}^{n} \Theta\left(\lambda - E_j\left(\frac{2\pi m_1}{n}, \frac{2\pi m_2}{n}\right)\right)$$

$$= \frac{1}{4\pi^2 ab}\sum_{j=1}^{a \cdot b} \int_{R^2/2\pi Z^2} \Theta(\lambda - E_j(k))\,dk,$$

so that the density of states

$$\frac{d\rho}{d\lambda}(\lambda) = \frac{1}{4\pi^2 ab} \sum_{j=1}^{a \cdot b} \int_{R^2/2\pi Z^2} \delta(\lambda - E_j(k)) \, dk$$

$$= \frac{1}{4\pi^2 ab} \sum_{j=1}^{a \cdot b} \int_{E_j(k) - \lambda = 0} \frac{ds}{|\nabla_k E_j|},$$

upon evaluation of the delta function.

We may rewrite the last identity as

$$\frac{d\rho}{d\lambda}(\lambda) = \frac{1}{4\pi^2 ab} \int_{\bigcup_{j=1}^{ab} \{k \in R^2/2\pi Z^2 | E_j(k) - \lambda = 0\}} \frac{ds}{|\nabla_k E_j|}$$

$$= \frac{1}{4\pi^2 ab} \int_{F_\lambda \cap |\xi_1| = |\xi_2| = 1} \omega_\lambda,$$

where

$$\omega_\lambda = -\frac{P_\lambda \, d\xi_1}{\xi_1 \xi_2 P_{\xi_2}} = \frac{P_\lambda \, d\xi_2}{\xi_1 \xi_2 P_{\xi_2}}.$$

Here $P(\xi_1, \xi_2, \lambda)$ is the polynomial in ξ_1, $1/\xi_1$, ξ_2, $1/\xi_2$ and λ introduced in §2, which defines $B(V)$.

Let C be a compact algebraic curve and $\nu : \widehat{C} \to C$ its normalization. Recall that the dualizing sheaf K_C of C associates to each open set $U \subset C$ the set of meromorphic differential forms ω on $\nu^{-1}(U)$ such that for all $x \in U$ and all germs f of holomorphic functions on C at x one has

$$\sum_{y \in \nu^{-1}(x)} \operatorname{res}_y f\omega = 0.$$

Global sections of K_C are called Rosenlicht differentials on C. One has

arithmetic genus of $C = \dim_{\mathbb{C}} \Gamma(C, K_C)$.

All of the dualizing sheaves on the compactified Fermi curves F_λ can be put together to form the relative dualizing sheaf $K_{B(V)_{\text{comp}}/P^1}$ which has the property that its restriction to each of the curves F_λ is K_{F_λ}. At a smooth point p of $B(V)_{\text{comp}}$ where π has maximal rank the stalk of $K_{B(V)_{\text{comp}}/P^1}$ coincides with the quotient of the sheaf of holomorphic one forms on $B(V)_{\text{comp}}$ by the subsheaf generated by the forms $\pi^*(\eta)$ where η is a holomorphic one form on P^1 around p. A section of $K_{B(V)_{\text{comp}}/P^1}$ over a subset U of $B(V)_{\text{comp}}$ is called a relative differential form on U. If η is a relative differential form on $U \subset B(V)_{\text{comp}} \setminus F_\infty$ then $\eta \wedge \pi^*(d\lambda)$ is a section of $K_{B(V)_{\text{comp}}}$, the dualizing sheaf of $B(V)_{\text{comp}}$, and η is uniquely determined by it.

Observe that $(d\xi_1 \wedge d\xi_2)/(\xi_1 \xi_2)$ is a section of $K_{B(V)_{\text{comp}}}$ over $B(V)_{\text{aff}}$. We are interested in the section ω of $K_{B(V)_{\text{comp}}/P^1}$ characterized by

$$\omega \wedge \pi^*(d\lambda) = -\frac{d\xi_1 \wedge d\xi_2}{\xi_1 \xi_2}.$$

It is easy to see that the restriction of ω to F_λ is precisely the ω_λ we obtained above. We call ω the density of states form. Set

$$\beta := \left(\frac{\partial p}{\partial \lambda}\right)^{-1}\omega,$$

and let β_λ denote the restriction of β to F_λ.

THEOREM.

(1) For $\lambda \in \mathbb{C}$, ω_λ and β_λ are Rosenlicht differentials on the compactified Fermi curve F_λ. The form β_λ is nowhere vanishing on $F_{\mathrm{aff},\lambda}$ and vanishes with multiplicity $ab-1$ at each point of $F_\lambda \cap \Sigma^{e,f}$, $e,f = 0,\infty$.

(2) For generic $\lambda \in \mathbb{C}$ the form ω_λ on F_λ lifts to a holomorphic form on the normalization of F_λ.

(3) If $B(V)$ is smooth then $(d\xi_1 \wedge d\xi_2)/(\xi_1\xi_2\partial P/\partial\lambda)$ has no zeroes or poles on $B(V)$ but vanishes along each curve $\Sigma^{e,f}$ $(e,f = 0,\infty)$ on $B(V)_{\mathrm{comp}}$ with multiplicity $ab-1$.

(4) $\lambda \to \omega_\lambda$ defines a holomorphic section of $\pi_*(K_{B(V)_{\mathrm{comp}}/P^1})$ with a zero of order one at ∞.

(5) If the curves H_i^e $(i = 1,2, ; e = 0,\infty)$ are all smooth then the section $\lambda \to \lambda\cdot\omega_\lambda$ of $\pi_*(K_{B(V)_{\mathrm{comp}}/P^1})$ takes as value at ∞ a Rosenlicht differential ω_∞ on F_∞ with the following property: If $\nu: \hat{F}_\infty \to F_\infty$ is the normalization of F_∞ then $\nu^*(\omega_\infty)$ has poles only over the four points of $H_2^e \cap Q^{ef}$, $H_1^f \cap Q^{ef}$, $e,f = 0,\infty$, and the residue of $\nu^*(\omega_\infty)$ at these points is $\pm ab$.

Recall the covering map c of §3. By the above, the pull back $\tilde\omega = c^*(\omega)$ of the density of states form is a section of the relative dualizing sheaf of the lifted Bloch variety $\tilde\pi : \tilde{B}(V)_{\mathrm{comp}} \to P^1$ which defines a holomorphic differential form on a generic $\tilde{F}_\lambda$. Thus for $V = 0$ the density of states is given by

$$\frac{d\rho}{d\lambda} = \int_{E_\lambda \cap \{|\xi_1|=|\xi_2|=1\}} \tilde\omega_\lambda,$$

so that it is a period integral of the elliptic family $\pi: E \to P^1$. In particular it satisfies a Fuchsian differential equation. One can show that $\lambda(d\rho/d\lambda)$ is a solution of the hypergeometric differential equation*

$$P\left\{\begin{array}{ccc} 0 & 1 & \infty \\ \hline 0 & 0 & \frac{1}{2} \\ & & \\ 0 & 0 & \frac{1}{2} \end{array} \; t\right\}, \qquad t = \frac{\lambda^2-4}{\lambda^2}.$$

*We thank Chris Peters for helpful conversations.

7. Monodromy and density of states

We assume that V is a real potential that is generic in the sense of §4, that is $V \in \mathcal{E}$. For every real λ that is not a van Hove singularity the image of

$$\{(\xi_1, \xi_2, \lambda) \in F_\lambda | |\xi_1| = |\xi_2| = 1\},$$

under the quotient map $F_\lambda \to F_\lambda / \langle i \rangle = Y_\lambda$ defines a homology class $\alpha_\lambda \in H_1(Y_\lambda, Z)$. The density of states form ω is invariant under the involution i. Therefore it defines a relative differential form on Y which (by abuse of notation) we also call ω. By the observation of the previous section the density of states function $d\rho/d\lambda$ fulfills

$$\frac{d\rho}{d\lambda} = 2 \int_{\alpha_\lambda} \omega_\lambda, \quad \text{for all } \lambda \in R.$$

So $\frac{1}{2}(d\rho/d\lambda)$ is the period integral of the differential form ω_λ over the cycle α_λ. This shows that $d\rho/d\lambda$ is an analytic function of λ outside the set of van Hove singularities. At the real van Hove singularities one actually observes discontinuities of $d\rho/d\lambda$ or its derivative ([**AM**, Figure 8.3 and exercise 8.2]).

The purpose of this section is to compare the monodromy action on α_λ with the analytic continuation of the density of states function. For that purpose let $\mathcal{O}$ be the ring of holomorphic functions on the universal cover $\widetilde{P^1 \setminus D}$ of $P^1 \setminus D$. If λ_0 is a point of $P^1 \setminus D$ we define a map

$$I_{\lambda_0} : H_1(Y_{\lambda_0}, Z) \to \mathcal{O},$$

as follows: Let $\alpha \in H_1(Y_{\lambda_0}, Z)$ and $x \in \widetilde{P^1 \setminus D}$. The point x corresponds to the homotopy class of a path w in $P^1 \setminus D$ which starts at λ_0 and ends at a point $\lambda \in P^1 \setminus D$ lying under the point x. We put

$$I_{\lambda_0}(\alpha)(x) := \int_{r([w])(\alpha)} \omega_\lambda.$$

LEMMA. *For each* $\lambda_0 \in P^1 \setminus D$ *and each* $\alpha \in H_1(Y_{\lambda_0}, Z)$ *the function* $I_{\lambda_0}(\alpha)$ *is holomorphic on* $P^1 \setminus D$. *The map*

$$I_{\lambda_0} : H_1(Y_{\lambda_0}, Z) \to \mathcal{O},$$

is injective and equivariant with respect to the action of $\pi_1(P^1 \setminus D, \lambda_0)$. *Here, as before,* $\pi_1(P^1 \setminus D, \lambda_0)$ *acts on* $H_1(Y_{\lambda_0}, Z)$ *by the monodromy representation, and it acts on* $\mathcal{O}$ *as the covering group of* $\widetilde{P^1 \setminus D} \to P^1 \setminus D$.

All the statements of the lemma apart from the injectivity of I_{λ_0} are standard. Clearly the kernel of I_{λ_0} is a monodromy-invariant subspace. By

Theorem 1 in §5 it is either $\{0\}$ or $H_1(Y_{\lambda_0}, Z)$. The second case cannot occur because ω is a nonzero relative differential form.

The analytic continuation of the density of states function from a point $\lambda \in P^1 \setminus D$ is a function on $\widetilde{P^1 \setminus D}$ that lies in the image of I_λ. Let $\mathcal{O}'$ be the sublattice of $\mathcal{O}$ generated by all the $\pi_1(P^1 \setminus D, \lambda)$-images of this function. Via I_λ^{-1} this lattice is isomorphic to the sublattice of $H_1(Y_\lambda, Z)$ generated by the monodromy images of α_λ. However we would like to recover all of $I_\lambda(H_1(Y_\lambda, Z))$ from the analytic continuation of the density of states function. This is done by:

THEOREM 1. *Suppose that* $V \in \mathcal{E}$ *is a real-valued potential. Let* λ_0 *be a point on the negative real axis close to* $-\infty$, *and let* w_∞ *be a simple loop in* $P^1 \setminus D$ *starting and ending at* λ_0 *which separates* ∞ *from all other van Hove singularities. Let* α *be a nonzero element of* $H_1(Y_{\lambda_0}, Z)$ *and let* $\mathcal{O}'$ *be the sublattice of* $\mathcal{O}$ *generated by* $I_{\lambda_0}(\alpha)$ *and all its monodromy translates. Then for any two functions* $f_1, f_2 \in \mathcal{O}'$ *that are real (that is* $f_i(\overline{x}) = f_i(x)$ *where* $x \to \overline{x}$ *is the lifting of the complex conjugation) the functions*

$$r([w_\infty])(f_1) - f_1 \quad and \quad r([w_\infty])(f_2) - f_2,$$

are proportional. Furthermore there is a real f *in* $\mathcal{O}'$ *such that* $\tilde{g} := r([w_\beta])(f) - f$ *is not identically zero. Put*

$$g_\infty := \frac{2\pi i a b}{\lim\limits_{t \to -\infty} t g(t)} \cdot \tilde{g}.$$

Then $I_{\lambda_0}(H_1(Y_{\lambda_0}, Z))$ *is the sublattice of* $\mathcal{O}$ *generated by* g_∞ *and all the functions*

$$\frac{1}{2}(g_\infty - r([w])g_\infty), \qquad [w] \in \pi_1(P^1 \setminus D, \lambda_0).$$

The crucial point in the proof of Theorem 1 is to show that $g_\infty := I_{\lambda_0}(\gamma_\infty)$ (where $\gamma_\infty \in H_1(Y_{\lambda_0}, Z)$ is the cycle described in §5 can be characterized in the way described above.

Theorem 1 shows that one can describe the $\pi_1(P^1 \setminus D, \lambda_0)$-module $H_1(Y_{\lambda_0}, Z)$ in terms of the analytic continuation of the germ of the density of states function at a point $\lambda_* \in R$ where it is not identically zero (choose a path w joining λ_* to λ_0 in $P^1 \setminus D$ and take $\alpha := r([w]) \cdot \alpha_{\lambda_*}$ in the theorem). The density of states function can also be used to distinguish the spectral van Hove singularities from all the other van Hove singularities. This is achieved by the following theorem which is a direct consequence of the Picard-Lefschetz formulas.

THEOREM 2. *Let* $V \in \mathcal{E}$ *and* $\lambda_0 \in P^1 \setminus D$. *Choose a system* w_λ, $\lambda \in D$ *of simple loops starting and ending at* λ_0 *such that* w_λ *separates* λ *from all other points of* D, w_λ *and* $w_{\lambda'}$ *only intersect at* λ_0 *if* $\lambda \neq \lambda'$.

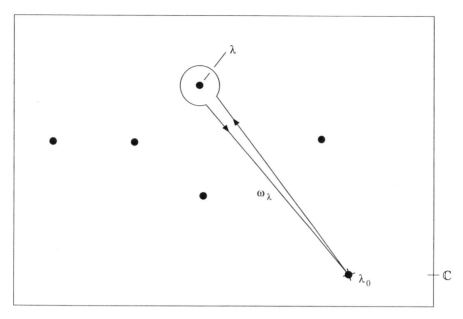

FIGURE 11. · denotes points of D.

For each $\lambda \in D \setminus \{\infty\}$ the set $\{r([w_\lambda]) \cdot f - f | f \in I_{\lambda_0}(H_1(Y_{\lambda_0}, Z))\}$ is a free Z-module of rank one. Let g_λ be a generator of this module. Then λ is a spectral van Hove singularity if and only if for all $\lambda' \in D - \{0\}$ there is $k \in Z$ such that

$$r([w_{\lambda'}]) \cdot g_\lambda - g_\lambda = 2k \cdot g_{\lambda'}.$$

Theorems 1 and 2 show that for real valued $V \in \mathscr{E}$ the density of states function determines the periodic-antiperiodic spectrum of the operator V. In the continuum case it is known that for a generic lattice Γ this information determines the Bloch variety (see [ERT §6]).

Finally, let us point out that the density of states function allows to determine whether V is generic in the sense of §4.

THEOREM 3. *Let V be a real-valued potential, and $\lambda_* \in R$ a point where the density of states function $d\rho/d\lambda$ is analytic and not identically zero in a neighborhood of λ_*. Then $V \in \mathscr{E}$ if and only if*

   (i) *the analytic continuation of the germ of $d\rho/d\lambda$ at λ_* has precisely*
       $\nu_{ab} = 2a^2b^2 - 6(a^2 + b^2) + 36(a + b) + 1$ *ramification points, and,*
   (ii) *if D is the set of ramification points then the $\pi_1(P^1 \setminus D, \lambda_0)$-orbit of the analytic continuation of the density of states function generates a sublattice of $\mathscr{O}$ of rank $a \cdot b$.*

This theorem can be proved as follows. The rank of the lattice in $\mathscr{O}$ generated by the $\pi_1(P^1 \setminus D, \lambda_0)$-orbit of the analytic continuation of the density of states function is at most rank $H_1(Y_{\lambda_0}, Z)$. If $V \in \mathscr{E}$ then equality holds by Theorem 1. Conversely, if this lattice has rank $a \cdot b$ then Y_{λ_0}

is smooth. Then the condition (i) of the theorem ensures that $\pi : Y(V) \to P^1$ has precisely $\nu_{a,b}$ singular fibers. Computing various Euler characteristics one sees that all singularities of π are of the nature described in §4.

8. The density of states determines the Bloch variety

Theorem 2 in §2 states that generically the Bloch variety $B(V)$ determines the potential V up to the obvious symmetries. Below we will see that generically the density of states function determines the Bloch variety. Together these two results show that for a generic potential the physically measurable quantity $d\rho/d\lambda$ determines the potential V up to symmetries.

THEOREM 1. *Let V, $V' \in L^2(Z^2/aZ \oplus bZ)$ be two real-valued potentials, and let $\lambda_* \in R$. Assume that the density of states functions $d\rho/d\lambda$ resp. $d\rho'/d\lambda$ of V resp. V' coincide in a neighborhood of λ_* and fulfill the conditions of Theorem 3 of §7. (This is the case for generic V, V'.) Then*

$$B(V) = B(V') \quad or \quad B(V) = i_1 B(V'),$$

where $i_1 : \mathbb{C}^* \times \mathbb{C}^* \times \mathbb{C} \to \mathbb{C}^* \times \mathbb{C}^* \times \mathbb{C}$ is the involution $(\xi_1, \xi_2, \lambda) \to (\xi_1^{-1}, \xi_2, \lambda)$.

We sketch the proof of this theorem. By shifting λ_* a little we may assume that λ_* is not a van Hove singularity for V or V'. From Theorem 3 of §7 it follows that V, $V' \in \mathscr{E}$. In §7 we also explained that the set of van Hove singularities of V resp. V' is precisely the set of ramification points of the analytic continuation of $d\rho/d\lambda$ resp. $d\rho'/d\lambda$. Therefore these two sets coincide, we call the set D. Now consider the maps

$$I_{\lambda_*} : H_1(Y(V)_{\lambda_*}, Z) \to \mathscr{O},$$

$$I'_{\lambda_*} : H_1(Y(V')_{\lambda_*}, Z) \to \mathscr{O},$$

defined in §7. By Theorem 1 of §7 the images of I_{λ_*} and I'_{λ_*} coincide. Since I_{λ_*} and I'_{λ_*} are both injective and equivariant with respect to the action of $\pi_1(P^1 \setminus D, \lambda_*)$

$$\varphi_* := (I'_{\lambda_*})^{-1} \circ I_{\lambda_*} : H_1(Y(V)_{\lambda_*}, Z) \to H_1(Y(V')_{\lambda_*}, Z),$$

is a monodromy equivariant isomorphism. From the Picard-Lefschetz formulas one concludes that φ_* also preserves the intersection form on these two spaces.

This can be phrased slightly differently. Let $R^1\pi_* Z$ resp. $R^1\pi'_* Z$ be the local systems over $P^1 \setminus D$ associated to $\pi : Y(V) \to P^1$ resp. $\pi' : Y(V') \to P^1$. The stalk of $R^1\pi_* Z$ resp. $R^1\pi'_* Z$ over a point $\lambda \in P^1 \setminus D$ is the cohomology $H^1(Y(V)_\lambda, Z)$ resp. $H^1(Y(V')_\lambda, Z)$. Since φ_* is monodromy equivariant it induces an isomorphism $\varphi^* : R^1\pi'_* Z \to R^1\pi_* Z$ of local systems. For each $\lambda \in P^1 \setminus D$ the restriction $\varphi^*_\lambda : H^1(Y(V')_\lambda, Z) \to H^1(Y(V)_\lambda, Z)$ preserves the intersection form on cohomology. The cohomology $H^1(Y(V)_\lambda, \mathbb{C})$ with complex coefficients can be decomposed into the direct sum of the spaces $H^{1,0}(Y(V)_\lambda)$ of cohomology classes representable by holomorphic 1-forms, and $H^{0,1}(Y(V)_\lambda)$, the space of classes representable by antiholomorphic 1-forms. The decomposition $H^1(Y(V)_\lambda, \mathbb{C}) = H^{1,0}(Y(V)_\lambda) \oplus H^{0,1}(Y(V)_\lambda)$ is called the *Hodge decomposition*. Similarly we have the Hodge decomposition $H^1(Y(V')_\lambda, \mathbb{C}) = H^{1,0}(Y(V')_\lambda) + H^{0,1}(Y(V')_\lambda)$. These decompositions make $R^1\pi_*Z$ and $R^1\pi'_*Z$ to algebraic families of polarized Hodge structures over $P^1 \setminus D$ in the sense of [**D**, 4.2]. The first basic step in the proof is:

LEMMA 1. φ^* *is an isomorphism of families of polarized Hodge structures. In other words, for each* $\lambda \in P^1 \setminus D$ *the map* φ^*_λ *is compatible with the Hodge decomposition*:
$$\varphi^*_\lambda(H^{1,0}(Y(V')_\lambda)) = H^{1,0}(Y(V)_\lambda) \quad and$$
$$\varphi^*_\lambda(H^{0,1}(Y(V')_\lambda)) = H^{0,1}(Y(V)_\lambda).$$

PROOF OF LEMMA 1. $\mathrm{Hom}_{P^1\setminus D}(R^1\pi'_*Z, R^1\pi_*Z)$ is also an algebraic family of polarized Hodge structures, and these Hodge structures have weight 0. Let $\mathrm{Hom}(R^1\pi'_*Z, R^1\pi_*Z)^0$ be the biggest constant subsystem. Its fiber is $H^0(P^1 \setminus D, \mathrm{Hom}_{P^1\setminus D}(R^1\pi'_*Z, R^1\pi_*Z))$, so it is isomorphic to the space of monodromy equivariant homomorphisms from $H^1(Y(V')_\lambda, Z)$ to $H^1(Y(V)_\lambda, Z)$. These two $\pi_1(P^1 \setminus D, \lambda_*)$-modules are isomorphic and the action of $\pi_1(P^1 \setminus D, \lambda_*)$ on them is absolutely irreducible (see §5). So by Schur's lemma $\mathrm{Hom}(R^1\pi'_*Z, R^1\pi_*Z)^0$ is a local system of rank one. Clearly φ^* is a generating section of this system.

By the theorem of the fixed part in [**D**, 4.1.2] the natural Hodge structure on $\mathrm{Hom}(R^1\pi'_*Z, R^1\pi_*Z)$ induces a Hodge structure on $\mathrm{Hom}(R^1\pi'_*Z, R^1\pi_*Z)^0$. This Hodge structure is of weight zero. Since the system has rank one, the induced Hodge structure is of type $(0, 0)$. This means that all elements of $\mathrm{Hom}(R^1\pi'_*Z, R^1\pi_*Z)^0$ are morphisms of Hodge structures. In particular this is true for φ^*.

The second main step is to apply the Torelli theorem for compact algebraic curves (see e.g., [**GH**, Chapter 2.7]). It shows that for each $\lambda \in P^{-1} \setminus D$ there is an isomorphism $\varphi_\lambda : Y(V)_\lambda \to Y(V')_\lambda$ that induces φ^*_λ on cohomology. One can put all these isomorphisms together to get

LEMMA 2. *There is an isomorphism* $\Phi : Y(V) \to Y(V')$ *such that the diagram*

$$Y(V) \xrightarrow{\ \Phi\ } Y(V')$$

$$\pi \searrow \qquad \swarrow \pi'$$

$$P^1$$

commutes, and such that the restriction of Φ *to* $Y(V)_\lambda$ *coincides with* φ_λ.

Then one analyzes the geometry of $Y(V)$ to see that Φ lifts to an isomorphism between $B(V)_{comp}$ and $B(V')_{comp}$. The linear system belonging to the embedding of $B(V)$ resp. $B(V')$ into $\mathbb{C}^* \times \mathbb{C}^* \times \mathbb{C}$ can be described in terms of the relative canonical system of $\pi : Y(V) \to P^1$ resp. $\pi' : Y(V') \to P^1$. This then makes it possible to show that the embeddings of $B(V)$ and $B(V')$ coincide up to the involution i_1.

REFERENCES

[AM] N. Ashcroft and N. Mermin, *Solid state physics*, Holt, New York, 1976.

[Bä] D. Bättig, *A toroidal compactification of the two-dimensional Bloch variety*, Thesis, ETH Zürich, 1988.

[Be] A. Beauville, *Les familles stables de courbes elliptiques sur P^1 admettant quatre fibres singulières*, C. R. Acad. Sci. Paris **294** (1982), 657–660.

[Bo] G. Borg, *Eine Umkehrung der Sturm-Liouvilleschen Eigenwertaufgabe*, Acta Math. **78** (1946), 1–96.

[D] P. Deligne, *Théorie de Hodge* 2, Inst. Hautes Études Sci. Publ. Math. **40** (1972), 5–57.

[ERT] G. Eskin, J. Ralston, and E. Trubowitz, *On isospectral periodic potentials in R^n*, Comm. Pure Appl. Math. **37** (1984), 647–676.

[GH] Ph. Griffiths, and J. Harris, *Principles of algebraic geometry*, Wiley, New York, 1978.

[KT] H. Knörrer and E. Trubowitz, *A directional compactification of the complex Bloch variety*, Comment. Math. Helvetici **65** (1990), 114–149.

[MT] H. McKean and E. Trubowitz, *Hills operator and hyperelliptic function theory in the presence of infinitely many branch points*, Comm. Pure Appl. Math. **29** (1976), 143–226.

[RS] M. Reed and B. Simon, *Methods of modern mathematical physics* IV, *Analysis of operators*, Academic Press, New York, 1978.

UNIVERSITY OF CALIFORNIA AT LOS ANGELES, LOS ANGELES, CA 90024

EIDGENÖSSISCHE TECHNISCHE HOCHSCHULE ZÜRICH, SWITZERLAND (H. KNÖRRER AND E. TRUBOWITZ)

Contemporary Mathematics
Volume **116**, 1991

Automorphisms of Cuspidal K3-Like Surfaces

BRIAN HARBOURNE

ABSTRACT. The class of cuspidal K3-like surfaces is defined, these being certain smooth, complete rational surfaces with a reduced irreducible cuspidal anticanonical divisor. Denote by $\widehat{A}(X)$ the group of automorphisms of X modulo those acting trivially on $\operatorname{Pic}(X)$. For cuspidal K3-like surfaces there is a well-behaved notion of $\widehat{A}$-relatively minimal model, this being a relatively minimal model among those cuspidal K3-like surfaces X for which $\widehat{A}(X)$ is infinite. In characteristics other than 2, 3 and 5, a complete determination of these $\widehat{A}$-relatively minimal models is made.

I. Introduction

Consider the problem of classifying smooth complete surfaces X over an algebraically closed field k whose group $\operatorname{Aut}(X)$ of biregular transformations (i.e., automorphisms) is finite. If M is a relatively minimal model of X (meaning that M is a relatively minimal model and that there is a birational regular map of X to M), then X is obtained by blowing up points (possibly infinitely near) of M, and there is an injection $G \to \operatorname{Aut}(X)$ of the subgroup G of $\operatorname{Aut}(M)$ which fixes each point blown up. Except in the case that $\mathbf{P}^2$ is a relatively minimal model of X, the injection $G \to \operatorname{Aut}(X)$ is, up to finite groups, an equality. (Let us say that groups F and H are related if there is a homomorphism $h \colon F \to H$ for which $\ker(h)$ is finite and $h(F)$ has finite index in H. This relation generates equivalence *up to finite groups* on the category of groups.) In particular, $\operatorname{Aut}(X)$ is finite if and only if G is, thus reducing the classification problem for such surfaces to minimal surfaces (where it has been worked out in case k is the complex field [N]) plus a fixed-point problem that is well-understood at least in concept.

1980 *Mathematics Subject Classification* (1985 *Revision*). Primary 14J50; Secondary 14J26, 14E30.

Key words and phrases. automorphism group, rational, surface, algebraic surface, relatively minimal model, minimal model.

Partially supported by an NSF grant DMS-8601743.

This paper is in final form and no version of it will be submitted for publication elsewhere.

But say X is a *basic surface*, by which we mean $\mathbf{P}^2$ is a relatively minimal model of X. Then there need be no connection between G and $\mathrm{Aut}(X)$ (beyond the fact that G is a subgroup of $\mathrm{Aut}(X)$). In particular, G may be finite while $\mathrm{Aut}(X)$ is not, and the question of how such surfaces X arise with $\mathrm{Aut}(X)$ infinite is not understood.

In this paper we answer this question for a special class of basic surfaces over algebraically closed fields k of characteristics other than 2, 3 and 5. To be more explicit, let X be a basic surface. Then X is obtained by blowing up points $p_1, \ldots, p_n$ of $\mathbf{P}^2$ (possibly infinitely near) and $\mathrm{Pic}(X)$ is the free abelian group on the divisor classes $e_0, \ldots, e_n$, where e_0 is the class of the pullback of a line in $\mathbf{P}^2$ and $e_1, \ldots, e_n$ are the classes of the total transforms $E_1, \ldots, E_n$ of the points $p_1, \ldots, p_n$; the *Picard number* $n + 1$ of X is the rank of $\mathrm{Pic}(X)$. We obtain a homomorphism $\widehat{}: \mathrm{Aut}(X) \to \mathrm{GL}(\mathrm{Pic}(X))$ by sending $\alpha \in \mathrm{Aut}(X)$ to $\widehat{\alpha} = (\alpha^*)^{-1}$, where α^* is the automorphism induced on $\mathrm{Pic}(X)$ by pullback by α. We denote the image of $\widehat{}$ by $\widehat{\mathrm{A}}(X)$ and its kernel by $\mathrm{A}_0(X)$—or simply by $\widehat{\mathrm{A}}$ and A_0 when no confusion will arise. Of course, A_0 consists of automorphisms of X which lift from automorphisms of $\mathbf{P}^2$ (and thus $\mathrm{A}_0(X)$ is just what we referred to above as G). Hence it is $\widehat{\mathrm{A}}$ which is most mysterious and with which we will mostly be concerned.

Let us call a basic surface X *anticanonical* if X supports a reduced irreducible curve C whose divisor class $[C] \in \mathrm{Pic}(X)$ is the anticanonical class $-K_X$. In particular, X is anticanonical if it is obtained by blowing up smooth points of a reduced irreducible plane cubic curve. Denoting by $\pi: C \subset X$ the inclusion, we have the restriction homomorphism $\pi^*: \mathrm{Pic}(X) \to \mathrm{Pic}(C)$ and the degree homomorphism $d: a \in \mathrm{Pic}(X) \mapsto -a \cdot K_X \in \mathbf{Z}$. We call an anticanonical surface X K3-*like* if the group $\pi^*(\ker(d)) = \ker(d)/(\ker(d) \cap \ker(\pi^*))$ is finite. For example, any anticanonical surface over the algebraic closure of a finite field is K3-like; K3-like surfaces have an intricate and interesting biregular geometry suggestive of K3 surfaces [H1] (but note that the present definition of K3-like is slightly broader than that of [H1], where we required that a K3-like surface have Picard number at least 11). Of particular note here is that K3-like surfaces have a notion of $\widehat{\mathrm{A}}$-*relatively minimal model*, by which we mean relatively minimal among K3-like surfaces for which $\widehat{\mathrm{A}}$ is infinite:

THEOREM I.1. *Let $X \to Y$ be a birational morphism of* K3-*like surfaces. Then $\widehat{\mathrm{A}}(X)$ is infinite if $\widehat{\mathrm{A}}(Y)$ is.*

PROOF. This follows from [H1, lemma 3.7]. $\square$

Thus to classify the K3-like surfaces with infinite $\widehat{\mathrm{A}}$, it suffices to classify $\widehat{\mathrm{A}}$- relatively minimal models, concerning which we have:

CONJECTURE I.2 [H1]. *The $\widehat{\mathrm{A}}$-relatively minimal models are the K3-like surfaces of Picard number* 10 *with infinite $\widehat{\mathrm{A}}$.*

I.e., if X is K3-like and $\widehat{A}(X)$ is infinite, then it is conjectured that there is a birational morphism $X \to Y$ where Y has Picard number 10 and $\widehat{A}(Y)$ is infinite; the morphism $X \to Y$ ensures that Y is K3-like. If X is K3-like and the anticanonical curve C is a rational curve with a cusp (as opposed to C being smooth or having a node), then we call X *cuspidal*. This paper's major result, Theorem I.3, is a special case of the conjecture above:

THEOREM I.3. *In characteristics other than 2, 3 or 5, every cuspidal K3-like surface with infinite $\widehat{A}$h as an $\widehat{A}$-minimal model which is a cuspidal K3-like surface of Picard number 10.*

Conceptually, it is helpful to understand what exactly are the cuspidal K3-like surfaces of Picard number 10 with infinite $\widehat{A}$, and so we have:

THEOREM I.4. *Cuspidal K3-like surfaces of Picard number 10 with infinite $\widehat{A}$ occur only in positive characteristics, and in any positive characteristic greater than 5 they are precisely the rational minimal elliptic surfaces with an anticanonical or multiple fiber of type II but no fiber of type II*.*

In explanation, an *elliptic surface* is a smooth surface which is fibered in curves of arithmetic genus 1 and such that the general fiber is smooth. If the general fiber is not smooth, the surface is called *quasielliptic*, which is a phenomenon only of characteristics 2 and 3. We employ the term *numerically elliptic* to mean either elliptic or quasielliptic. *Minimal* here means that no fiber contains an exceptional curve, and we say an elliptic surface is *Jacobian* if the fibration has a section. The *fiber types* come from Kodaira's classification of the fibers of a numerically elliptic fibration (see [**BM**] and [**K**]). We note if the set of sections of the fibration of a rational minimal Jacobian numerically elliptic surface meet some irreducible fiber in only a finite number of points (and this is always the case over a ground field k which is the algebraic closure of a finite field), then the surface is a K3-like surface.

Finally, Corollary II.9 gives in all characteristics (including 0) other than 2, 3 and 5, a completely explicit description of the cuspidal K3-like surfaces with $\widehat{A}$ finite. The proofs of Theorems I.3 and I.4 are given in §II, while §III gives the proof of the main technical result on which Theorem I.3 is based.

In closing, we thank B. Goddard and L. Chouinard for very helpful conversations and for bringing to our attention useful facts, and the Spring 1988 UNL algebra seminarists for letting us work some of this material out in front of them.

II. Proofs of the theorems

Unless explicitly stated otherwise, we work over an algebraically closed ground field k of arbitrary characteristic. Now consider any sequence $X = X_{n+1} \to X_n \to \cdots \to X_1 = \mathbf{P}^2$ of blowings-up, where the morphism $X_{i+1} \to X_i$ is the monoidal transformation centered at a point p_i of X_i. Loosely,

X is the blowing up of (possibly infinitely near) points $p_1, \ldots, p_n$ of $\mathbf{P}^2$. As noted above, $\mathrm{Pic}(X)$ is the free abelian group on $e_0, \ldots, e_n$; moreover, the intersection product is obtained by linearity from: $e_i \cdot e_j = 0$, $i \neq j$; $e_i \cdot e_i = -1$, $i > 0$; and $e_0 \cdot e_0 = 1$. Such a basis of $\mathrm{Pic}(X)$ obtained from a birational morphism of X to $\mathbf{P}^2$ we call an *exceptional configuration*. The canonical class K_X of X is $-3e_0 + e_1 + \cdots + e_n$. The subgroup of $\mathrm{Pic}(X)$ orthogonal to K_X is denoted $K_X^\perp$. If $\pi: C \subset X$ is an anticanonical curve, then the kernel of $\pi^*: \mathrm{Pic}(X) \to \mathrm{Pic}(C)$ lies in $K_X^\perp$. The condition that X be K3-like is just that X have a reduced irreducible anticanonical curve $\pi: C \subset X$ such that $K_X^\perp / \ker(\pi^*)$ is finite.

Now $K_X^\perp$ is freely generated by the classes $r_0 = e_0 - e_1 - e_2 - e_3$, $r_1 = e_1 - e_2, \ldots, r_{n-1} = e_{n-1} - e_n$. One easily checks the well-known facts that the involutions $s_i: \mathrm{Pic}(X) \to \mathrm{Pic}(X)$ given by $s_i(x) = x + (x \cdot r_i)r_i$ preserve K_X and the intersection product (i.e., $s_i(x) \cdot s_i(y) = x \cdot y$) and hence $K_X^\perp$. Denote by W_n the subgroup of $\mathrm{GL}(\mathrm{Pic}(X))$ generated by $s_0, \ldots, s_{n-1}$.

PROPOSITION II.1.

(a) $\widehat{A}(X) \subset W_n$
(b) W_n *is infinite if and only if* $n \geq 9$.

PROOF. This is well-known; see [H1] and [H3] for discussion and more references.   □

LEMMA II.2. *Let* X *be K3-like.*

(a) *If* $X \to Y$ *is a birational morphism to a basic surface* Y, *then* Y *is K3-like; moreover,* Y *is cuspidal K3-like if* X *is.*
(b) *If* X *has Picard number* 10 *or more, then there exists a birational morphism* $X \to Y$ *where* Y *is a rational minimal numerically elliptic surface, and which is an isomorphism if* X *has Picard number* 10.

PROOF. (a) Any birational morphism of surfaces factors through a sequence of blowings up at smooth points, so X is obtained by blowing up points of Y. But if E is an exceptional curve and C is in the class of $-K_X$, then $E \cdot C = 1$ so the morphism $X \to Y$ restricts to an isomorphism of the anticanonical curve C of X (which exists since X is K3-like) to its image (which we find it convenient to also denote by C) which is itself anticanonical. From $C \subset X \to Y$ we see that the natural homomorphism $\mathrm{Pic}(Y) \to \mathrm{Pic}(C)$ factors through $\mathrm{Pic}(X)$ via the injection $\mathrm{Pic}(Y) \to \mathrm{Pic}(X)$, so X being K3-like and Y being basic forces Y to be K3-like too. And if X is cuspidal K3-like, then C is cuspidal, so Y is cuspidal K3-like.

(b) The Picard number of X is $n + 1$, where n is the number of points $p_1, \ldots, p_n$ of $\mathbf{P}^2$ blown up to obtain X, and by hypothesis $n \geq 9$. Let Y be the blowing up of $p_1, \ldots, p_9$ and let $X \to Y$ be the blowing up of the remaining points p_i. Since X is K3-like, there is an integral anticanonical curve $C \subset X$. By (a) and its proof, the image (again denoted C) of C in Y is an anticanonical curve with respect to which Y is also K3-like. Thus $K_Y \otimes \mathcal{O}_C$ is a torsion element of $\text{Pic}(C)$, so by (for example) [HL, Lemma 4.3] Y is minimal numerically elliptic. But every rational minimal numerically elliptic surface has Picard number 10 (cf., [HL]) so $X \to Y$ is an isomorphism if X also has Picard number 10. □

DEFINITION II.3. A rational minimal numerically elliptic surface Y, Jacobian or not, is said to be *extremal* if the sublattice Γ of $\text{Pic}(Y)$ generated by the divisor classes of components of fibers has finite index in $K_Y^{\perp}$ (or, equivalently, if the rank of the span in $\text{Pic}(Y)$ of the (-2)-curves of Y is 9).

NOTE II.4. It follows from the definition that every extremal surface with a reduced irreducible anticanonical curve is K3-like. But even though by Lemma II.2 a K3-like surface of Picard number 10 is numerically elliptic, it need not be extremal.

PROPOSITION II.5. *Let Y be a K3-like surface of Picard number 10 (and hence Y is a rational minimal numerically elliptic surface). Then $\widehat{A}(Y)$ is finite if and only if Y is extremal.*

PROOF. By [H1, Lemma 3.7], $\widehat{A}(Y)$ is finite if and only if Y has only finitely many exceptional configurations. But a rational minimal numerically elliptic surface has only finitely many exceptional configurations if and only if it has only finitely many (-1)-curves (i.e., irreducible exceptional curves of the first kind). (To see this, note that if E is an exceptional curve on Y, then there is a birational morphism of Y to $\mathbf{P}^2$ contracting E to a point [HL, Lemma 4.2]. Conversely, the prime divisors contracted in any birational morphism $Y \to \mathbf{P}^2$ are curves of negative self-intersection, and by the adjunction formula this self-intersection is -1 or -2. But a rational minimal numerically elliptic surface has only finitely many curves of self-intersection -2, since these are components of the reducible fibers of the numerically elliptic fibration on Y.) But Y has only finitely many (-1)-curves if and only if Y is extremal [HM, Lemma I.4 and Proposition I.5]. □

The reason that the main result of this paper is restricted to characteristics other than 2, 3 and 5 is that the analysis required is much simplified by the next fact.

PROPOSITION II.6. *Let Y be a rational minimal numerically elliptic surface with an integral anticanonical curve C having a cusp. In characteristics other than 2, 3 and 5, Y is extremal if and only if it has a fiber of Kodaira type II^*.*

PROOF. In any characteristic, whether or not C has a cusp, if Y has a type II^* fiber, then Y is extremal, since the classes in $\text{Pic}(Y)$ of the components of a reducible fiber are linearly independent (see, for example, the proof of [HL, Lemma 7.4]) and a type II^* fiber has 9 components.

Conversely, let L denote the subgroup of $K_Y^\perp$ generated by the components of reducible fibers of Y (i.e., L is generated by the (-2)-curves on Y). Now $K_Y^\perp/L$ is finite by extremality and by [HM, Theorem IV.5] $K_Y^\perp/L$ is isomorphic to a subgroup of $\mathrm{Pic}(C)$. Since C is cuspidal, $K_Y^\perp/L$ is either trivial (in characteristic 0) or (in positive characteristics) a direct sum of a finite number of copies of $\mathbf{Z}_p$, where p is the characteristic. But if Y were an extremal surface which does not have a fiber of type II^*, then $K_Y^\perp/L$ would have 2, 3 or 5 torsion since, up to a factor of p (which would result if C is a multiple fiber), the order of $K_Y^\perp/L$ is the square root of the product of the discriminants of the reducible fibers, and, apart from type II^* fibers, the discriminants of reducible fibers occurring on extremal surfaces are always products of 2, 3, and 5 (see [HM] for details). □

COROLLARY II.7. *There is up to isomorphism a unique Jacobian extremal rational minimal numerically elliptic surface that has both a type II^* fiber and an integral anticanonical cuspidal curve C. In characteristics other than 0, there is also a unique non-Jacobian extremal rational minimal numerically elliptic surface that has both a type II^* fiber and an integral anticanonical cuspidal curve C. We denote these surfaces by Y_J and Y_N, respectively.*

PROOF. Let Y be an extremal rational minimal numerically elliptic surface that has both a type II^* fiber and an integral anticanonical cuspidal curve C. First consider the case that Y is Jacobian. The dual graph of the type II^* fiber is given in ($*$).

$(*)$

(Nodes represent irreducible components of the fiber. The notation c_i in the diagram, standing for *component i*, is just for reference. The number of edges between distinct nodes c_i and c_j equals the intersection product $c_i \cdot c_j$ of the corresponding components, here always either 0 or 1.)

By [HM], Y has a unique exceptional curve E, and E meets only component c_8 of the II^* fiber, and it does so transversely. Contracting E makes c_8 an exceptional curve, whose contraction makes c_7 an exceptional curve, and so on down to c_1. Thus we construct a birational morphism $Y \to \mathbf{P}^2$, under which the component c_0 maps to a line. Reinterpreted, this means that Y is obtained by blowing up $\mathbf{P}^2$ at a smooth flex point p_1 of C, and then at the point p_2 of C infinitely near to p_1, and so on up to p_9. As long as the characteristic is not 3, there is up to isomorphism a unique cuspidal cubic in $\mathbf{P}^2$ and it has a unique flex, so Y is unique. If the characteristic is 3, there is at least a unique cuspidal cubic with a flex, although in this case every point is a flex. (We note that in characteristic 3 there also occur cuspidal cubics with no flexes; for example, $x^2z + xy^2 + y^3$.) But if C_1 and C_2 are both cuspidal cubics with a flex, on each of which we have chosen a particular flex, there is an automorphism of $\mathbf{P}^2$ taking C_1 and its chosen flex to C_2 and its chosen flex. So Y is unique in characteristic 3 too.

Uniqueness is settled; consider existence. If S is the 9-time consecutive blowing up of a flex point of a cuspidal plane cubic C, then $-K_S$ restricts trivially to C and C is anticanonical, so by [HL, Lemma 4.3 and Proposition 4.1] and [H4, Proposition 1.2(b)] S is a Jacobian numerically elliptic surface and the classes r_i, $i = 0, \ldots, 8$ defined in the second paragraph of §II are the components of a type II^* fiber.

Now say Y is a non-Jacobian extremal rational minimal numerically elliptic surface that has both a type II^* fiber and an integral anticanonical cuspidal curve C. (Note that simply connected multiple fibers are creatures of positive characteristics, so Y_N does not occur in characteristic 0; see, e.g., [HL].) By [HM], there are disjoint exceptional curves which we will denote E_9 and E_8, such that of the components c_i, $i = 0, \ldots, 8$, E_9 meets only component c_8, and E_8 meets only c_7 and c_8, the meeting with c_7 being transverse. Under the contraction of E_9 and E_8, the image of c_7 is exceptional and can be contracted, which results in c_6 being exceptional and so it too can be contracted. Continuing in this way, we contract $c_5, \ldots, c_1$, thereby obtaining a birational morphism $Y \to \mathbf{P}^2$. Since $c_1, \ldots, c_7$, E_8 are connected, we see that their image is a single point f of (the image of) C which is itself a cuspidal cubic curve in $\mathbf{P}^2$. Since c_0 maps to a line meeting C at the single point f, f must be a flex. If we denote the image of E_9 by q, then we see that Y is obtained by blowing up a flex of C eight times and some other smooth point q once. But as above there is up to automorphisms of $\mathbf{P}^2$ a unique plane cuspidal curve with flex. In appropriate homogeneous coordinates on $\mathbf{P}^2$, this curve C can be defined by $y^3 - x^2z = 0$, with $[1 : 0 : 0]$ being the flex f. But for any nonzero constant t in the ground field k, $[x : y : z] \to [x : ty : t^3z]$ gives an automorphism of $\mathbf{P}^2$ preserving C and fixing f, and, for appropriate choice of t, moving q to any other desired smooth point of C. Thus Y is unique up to isomorphism.

To check existence (and now we may assume the characteristic p is positive), let S be the blowing up of a cuspidal plane cubic C eight times at the flex and once at any other smooth point of C. Then by direct calculation, keeping in mind that C is of additive type, $K_S \otimes \mathcal{O}_C$ is a torsion element of $\mathrm{Pic}(C)$ of order p, so by (for example) [HL, Lemma 4.3], S is minimal numerically elliptic, C is anticanonical and pC is a multiple fiber. Again one sees directly that $r_0, \ldots, r_7$ are the classes of (-2)-curves and hence of components of fibers. Their intersection diagram D is connected, comprising nodes $c_0, \ldots, c_7$ of diagram $(*)$ above. But by a perusal of the Kodaira classification of fibers of numerically elliptic fibrations [K, BM] we see that the only intersection diagram of a fiber which contains D is that of a fiber of type II^*.   □

The following proposition is our main technical result. We defer its proof to §III.

PROPOSITION II.8. *Let X be cuspidal K3-like in positive characteristic.*

  (a) *If $\hat{A}(X)$ is infinite and there is a nonisomorphic birational morphism $X \to Y_J$, then there is a birational morphism $X \to Y_N$.*

  (b) *If the characteristic is at least 7 and if there is a nonisomorphic birational morphism $X \to Y_N$, then there is a birational morphism $X \to Z$ where Z is a cuspidal K3-like surface of Picard number 10 with $\hat{A}(Z)$ infinite, and hence $\hat{A}(X)$ is infinite.*

We can now prove Theorem I.3.

PROOF OF THEOREM I.3. The Picard number of $\hat{A}$-relatively minimal models is at least 10 by Proposition II.1, so if X is cuspidal K3-like with Picard number more than 10 and if $\hat{A}(X)$ is infinite, we must show that X is not $\hat{A}$-relatively minimal. By Lemma II.2 there is a birational morphism $X \to Y$ with Y numerically elliptic and cuspidal K3-like. If $\hat{A}(Y)$ is infinite then X is not $\hat{A}$-relatively minimal and we are done. So assume $\hat{A}(Y)$ is finite. Then Y is extremal by Proposition II.5 and so has a type II^* fiber by Proposition II.6 and hence Y is either Y_J or Y_N by Corollary II.7. By Proposition II.8(a), there is a nonisomorphic birational morphism $X \to Y_N$, whence by Proposition II.8(b) there is a nonisomorphic birational morphism $X \to Z$ with $\hat{A}(Z)$ infinite, and so X is not $\hat{A}$-relatively minimal.   □

Using Proposition II.8 we can also classify the cuspidal K3-like surfaces with finite $\hat{A}$ in characteristics other than 2, 3 or 5.

COROLLARY II.9. *Let X be cuspidally K3-like of Picard number $n+1$ and let the characteristic not be 2, 3 or 5. Then $\hat{A}(X)$ is finite if and only if:*

  • *$n \leq 8$, or*
  • *X is obtained by the n-time consecutive blowing up of the flex of a cuspidal plane cubic curve, or*
  • *$X = Y_N$, $n = 9$ and the characteristic is (necessarily) not 0.*

PROOF. If $n \leq 8$, then $\hat{A}(X)$ is finite by Proposition II.1. So let X be obtained by the n-time consecutive blowing up of the flex of a cuspidal plane cubic curve. Then with respect to these blowings up the classes $r_0, \ldots, r_{n-1}$ (viz. paragraph 2 of §II) are the classes of (-2)-curves on X and these classes generate $K_X^\perp$. But (-2)-curves are irreducible, the class of any (-2)-curve is a nonnegative integer linear combination of $r_0, \ldots, r_{n-1}$ [H5, Lemma (0.2)], and the class of a (-2)-curve has only one effective representative, so $r_0, \ldots, r_{n-1}$ are the only classes of (-2)-curves. So $\hat{A}(X)$ is finite by [H1, Theorem 3.2]. Finally, if $X = Y_N$ then X is extremal elliptic with a cuspidal anticanonical curve, so, by [H3, Proposition 3.2(1)] and by [H3, Theorem (4.1)(2)] and its proof, $\hat{A}(X)$ is finite.

Conversely, say $\widehat{A}(X)$ is finite with $n > 8$. If $n = 9$, then X is numerically elliptic by Lemma II.2, extremal by Proposition II.5, with a type II^* fiber by Proposition II.6. By Corollary II.7, X is either Y_N and the characteristic is positive or X is Y_J. But by the proof of Corollary II.7 the latter is obtained by the 9-time consecutive blowing up at the flex of a cuspidal plane cubic curve C. If $n \geq 10$, then X is; by Lemma II.2, Theorem I.1, and Propositions II.5, and II.6, and Corollary II.7, a blowing up of a surface Y which is either Y_J or Y_N. By Proposition II.8, Y has to be Y_J; i.e., Y is obtained by successive blowings up at the flex of a cuspidal plane cubic curve C. But in characteristic 0, any blowing up S of Y_J away from the flex is not K3-like (since $\pi^*(K_S^{\perp})$ would not be trivial which forces it to be infinite since $\operatorname{Pic}^0(C)$ is torsion-free in characteristic 0), while in positive characteristics, by reordering the blowings up, any blowing up S of Y_J away from the flex is actually a blowing up of Y_N, as is clear from the constructions of Y_J and Y_N in the proof of Corollary II.7. But by Proposition II.8 this results in $\widehat{A}(X)$ being infinite, contrary to the hypothesis. Thus X must be obtained by blowing up the Jacobian extremal surface only at the flex.   □

PROOF OF THEOREM I.4. Let X be cuspidal K3-like and let C be a cuspidal anticanonical curve. Thus X is obtained by blowing up smooth points of a cuspidal cubic plane curve C. For X to be K3-like in characteristic 0 the blowings up must be consecutive blowings up of the flex of C, but by Corollary II.9 the surface X thus created has finite $\widehat{A}(X)$. So if $\widehat{A}(X)$ is infinite, then the characteristic p is positive.

Now if X has Picard number 10, it is rational minimal numerically elliptic by Lemma II.2 and C is a type II fiber, either anticanonical or multiple. No type II^* fiber can occur since if one does, X is extremal by Proposition II.6 and $\widehat{A}(X)$ is then finite by Proposition II.5.

Conversely, say X is rational minimal numerically elliptic (and hence a blowing up of $\mathbf{P}^2$ [HL, Lemma 4.2]) with a type II fiber that is either anticanonical or multiple. Then the support C of this fiber is a cuspidal anticanonical curve (which is clear if the fiber is anticanonical and which follows if the fiber is multiple since a multiple fiber on a rational surface is always a multiple of an anticanonical curve). But the group $\operatorname{Pic}^0(C)$ of degree zero divisor classes on C is pure p-torsion so the image of $K_X^{\perp}$ is finite and X is K3-like. And any cuspidal K3-like rational minimal numerically elliptic surface X in characteristics greater than 5 has by Propositions II.5 and II.6 infinite $\widehat{A}(X)$ if and only if it does not have a type II^* fiber.   □

REMARK II.10. We have here an amusing observation. If X is cuspidal K3-like with Picard number $n+1 \geq 11$ in some characteristic other than 2, 3 or 5, then either $\widehat{A}(X)$ or $A_0(X)$ is infinite (but never both [H2]) and hence $\operatorname{Aut}(X)$ is always infinite. For if $\widehat{A}(X)$ is finite, then by Corollary II.9 X is obtained by blowing up the flex of a cuspidal plane cubic n times. With respect to homogeneous coordinates $[x : y : z]$ on $\mathbf{P}^2$, we can take the cubic to be $y^3 - x^2 z = 0$ so the flex is $[1 : 0 : 0]$. Then for each nonzero element t of the ground field, the map $[x : y : z] \mapsto [x : ty : t^3 z]$ is in $\text{Aut}(\mathbf{P}^2)$ and lifts to an element $A_0(X)$ of X, so $\widehat{A}_0(X)$ is infinite. Moreover, if $\widehat{A}(X)$ is finite (and $n + 1 \geq 11$), then $\widehat{A}(X)$ is trivial since one can easily check that X has a unique exceptional configuration [H3].

III. The proof of the main technical result

PROOF OF PROPOSITION II.8. (a) Now Y_J is obtained by the 9-time consecutive blowing up of the flex point of a cuspidal plane cubic C, and $X \to Y_J$ is obtained by $n - 9$ more blowings up of smooth points of C, where $n + 1$ is the Picard number of X, n being greater than 9 since $X \to Y_J$ is nonisomorphic by hypothesis. If these extra $n - 9$ blowings up are all centered at the flex point of C, then $\widehat{A}(X)$ is finite, by Corollary II.9. So if we are to have $\widehat{A}(X)$ infinite, one of these last $n - 9$ blowings up must be centered away from the flex. If we reorder the blowings up so that the first eight, as before, are centered at the flex but the ninth is centered away from the flex, then by the construction of Y_N in the proof of Corollary II.7 the surface S resulting from the first 9 blowings up is Y_N, thereby giving the desired morphism $X \to Y_N$.

(b) It is enough to show that there is the required birational morphism $X \to Z$ for any blowing up $X \to Y_N$ at a single smooth point q of the cuspidal anticanonical curve C of Y_N. We now make a useful observation. Let Y be a cuspidal K3-like surface of Picard number 10 and denote by C a cuspidal anticanonical curve on Y; as always we have $\pi^*: \text{Pic}(Y) \to \text{Pic}(C)$. By Lemma II.2(b) and Propositions II.5 and II.6, $\widehat{A}(X)$ is finite if and only if Y is either Y_J or Y_N. But from the constructions of Corollary II.7 we see $K_Y^{\perp}/\ker(\pi^*)$ is $\{0\}$ in the former contingency and $\mathbf{Z}_p$ in the latter, where p is the characteristic. In particular, if $\pi^*(K_Y^{\perp})$ has order greater than p, then $\widehat{A}(X)$ is infinite.

Now let $e_1, \ldots, e_8 \in \text{Pic}(Y_N)$ be the classes of the exceptional curves $E_1, \ldots, E_8$ corresponding to the eight blowings up of C at its flex. The ninth blowing up is at a point p_9 of C away from both the flex and cusp of C. We denote the corresponding exceptional curve by E_9 and its class by e_9. If we denote the flex of C by f, then $\pi^*(K_{Y_N}^{\perp})$ is generated by $\mathcal{O}_C(p_9 - f)$. If q is a smooth point of C but $\mathcal{O}_C(q - f)$ is not a multiple of $\mathcal{O}_C(p_9 - f)$ in $\text{Pic}(C)$, then the surface Z obtained by blowing up q and contracting E_8 will have $\pi^*(K_Z^{\perp}) = \mathbf{Z}_p \oplus \mathbf{Z}_p$, generated by $\mathcal{O}_C(p_9 - f)$ and $\mathcal{O}_C(q - f)$, whence by our observation above $\widehat{A}(Z)$ is infinite as required.

So we only need to consider the case that $\mathcal{O}_C(q - f)$ is one of the p multiples of $\mathcal{O}_C(p_9 - f)$ in $\pi^*(K_{Y_N}^{\perp})$. Suppose that we can find an irreducible exceptional curve $E \neq E_9$ on Y_N such that the restriction to C of $[E] - K_{Y_N}$ is the divisor class in $\text{Pic}(C)$ of q. Then I claim that by blowing up q and contracting E, we obtain a birational morphism $X \to Z$ in which Z is a cuspidal K3-like surface of Picard number 10 with $\widehat{A}(Z)$ infinite as desired.

To justify the claim, note that since $\pi^*(K_{Y_N})$ is nontrivial E cannot meet C at q, so the contraction of E and blowing up of q can be done in either order. By blowing up q first we obtain X, and then by contracting E we obtain a birational morphism $X \to Z$. Clearly, Z has Picard number 10 and $\pi^*(K_Z^{\perp})$ is finite. Finally, Z is basic since it has an irreducible anticanonical curve, C. (The only smooth complete rational surfaces having a reduced irreducible anticanonical curve are $P^1 \times P^1$, the Hirzebruch surface F_2, both of which have Picard number 2, and blowings up of $\mathbf{P}^2$ at smooth points of an integral plane cubic.) Thus Z is cuspidal K3-like.

To see that $\widehat{A}(Z)$ is infinite, note that by construction K_Z restricts to C trivially. If we denote by T the image of the morphism $Y_N \to T$ contracting E, the image of q in T is the base point of the linear system $|-K_T|$; i.e., T is the blowing up of 8 of the nine base points (q being the ninth) of a plane cubic pencil. This makes Z a Jacobian elliptic surface, its elliptic fibration coming from the pencil $|-K_T|$ on T via pullback by the blowing up $Z \to T$ of q. The (-2)-curves on T are precisely the images under the morphism $Z \to T$ of the (-2)-curves on Z disjoint from the blowing up E_q of the point q. Since E_q is a section of the elliptic fibration on Z, the number of (-2)-curves on T is the sum $\sum_F (m_F - 1)$ over the reducible fibers F on Z, where m_F is the number of irreducible components of F. Hence, if Z were to have a type II^* fiber, then T would have at least eight (-2)-curves. On the other hand, the (-2)-curves on T are also the images under $Y_N \to T$ of the (-2)-curves on Y_N disjoint from E. But the only (-2)-curves on Y_N are components of the type II^* fiber (and thus correspond to nodes $c_0, \dots, c_8$ of the intersection diagram $(*)$ in section II). From [HM], in characteristics greater than 5, every exceptional curve on Y_N except E_9 meets (the components corresponding to) at least two of the nodes of $(*)$. Thus T has at most seven (-2)-curves, so Z cannot have a type II^* fiber, and hence we see $\widehat{A}(Z)$ is infinite (Propositions II.5 and II.6).

Let us amplify the claim that every exceptional curve on Y_N except E_9 meets at least two of the nodes of $(*)$. We have the exceptional classes $e_1, \dots, e_9$ on $\mathrm{Pic}(Y_N)$ and we have the pullback e_0 (with respect to the morphism $Y_N \to \mathbf{P}^2$ obtained by contracting $E_9, \dots, E_1$) of the class of a line. As in the second paragraph of §II we have the generators r_i, $i = 0, \dots, 8$ of $K_{Y_N}^{\perp}$. For $0 \le i \le 7$, the class r_i is the class of the component of the type II^* fiber of Y_N corresponding to node c_i of $(*)$. We will denote the class of the remaining component, which corresponds to c_8, simply by c. The span L of $\{r_i : i = 0, \dots, 7\}$ is isomorphic to the E_8 lattice (i.e., the root lattice of the E_8 lie algebra), except that L is negative definite rather than positive definite.

In particular, L is *unimodular* (i.e., the 8×8 matrix $[r_i \cdot r_j]$ has $\det([r_i \cdot r_j]) = 1$) and *even* (i.e., for $x, y \in L$, $x \cdot y$ is an even integer). Unimodularity implies that L has a dual basis $\check{r}_i$, $i = 0, \ldots, 7$ such that $\check{r}_i \cdot r_j$ is Kronecker's delta δ_{ij} (which is equal to 1 if $i = j$ and 0 otherwise).

Now [HM] finds that the classes of irreducible exceptional curves on Y_N are precisely the classes $e_9 + \check{\delta} + mK_{Y_N}$, where $m = \check{\delta} \cdot \check{\delta}/2 = \check{\delta}^2/2$ and $\check{\delta} = a_0 \check{r}_0 + \cdots + a_7 \check{r}_7$ such that a_i, $i = 0, \ldots, 7$ are nonnegative integers satisfying $3a_0 + 2a_1 + 4a_2 + 6a_3 + 5a_4 + 4a_5 + 3a_6 + 2a_7 \leq p$, p being the characteristic. From this it is easy to see that every exceptional curve on Y_N except E_9 meets r_i for some $i \leq 7$. Of course, we have $r_i \cdot c \geq 0$ for $i = 1, \ldots, 7$ since these correspond to distinct components of the type II^* fiber on Y_N.

The only solution to these inequalities is $c = r_8 + tK_{Y_N}$ for some integer t. But the fibers on Y_N are linearly equivalent to $-pK_{Y_N}$ (see, for example, the proof of Corollary II.7) so $t = 1 - p$. Now the 8×8 matrix $[\check{r}_i \cdot \check{r}_j]$ is just the inverse of $[r_i \cdot r_j]$, so by direct calculation we find $[\check{r}_i \cdot \check{r}_j]$ is the matrix $(**)$ displayed below.

Using $(**)$ we see that $(e_9 + \check{\delta} + mK_{Y_N}) \cdot c = p - (3a_0 + 2a_1 + 4a_2 + 6a_3 + 5a_4 + 4a_5 + 3a_6 + 2a_7)$. Thus if $e_9 + \check{\delta} + mK_{Y_N}$ is the class of an exceptional curve E on Y_N other than e_9 which meets r_i for at most one value, say $i = j$, of the index $i \leq 7$, then a_i is nonzero precisely for $i = j$. Since $p \geq 7$ is prime and $3a_0 + 2a_1 + 4a_2 + 6a_3 + 5a_4 + 4a_5 + 3a_6 + 2a_7 \leq p$, we see in fact that $3a_0 + 2a_1 + 4a_2 + 6a_3 + 5a_4 + 4a_5 + 3a_6 + 2a_7 < p$, whence E meets c and r_j; i.e., E meets at least two of the components of the type II^* fiber of Y_N, as claimed.

$$
(**) \quad [\check{r}_i \cdot \check{r}_j] =
\begin{bmatrix}
-8 & -5 & -10 & -15 & -12 & -9 & -6 & -3 \\
-5 & -4 & -7 & -10 & -8 & -6 & -4 & -2 \\
-10 & -7 & -14 & -20 & -16 & -12 & -8 & -4 \\
-15 & -10 & -20 & -30 & -24 & -18 & -12 & -6 \\
-12 & -8 & -16 & -24 & -20 & -15 & -10 & -5 \\
-9 & -6 & -12 & -18 & -15 & -12 & -8 & -4 \\
-6 & -4 & -8 & -12 & -10 & -8 & -6 & -3 \\
-3 & -2 & -4 & -6 & -5 & -4 & -3 & -2
\end{bmatrix}
$$

We are ready to translate the problem of finding an irreducible exceptional curve $E \neq E_9$ on Y_N such that the restriction to C of $[E] - K_{Y_N}$ is the divisor class in $\text{Pic}(C)$ of q. Now $\pi^*(K_{Y_N}^{\perp}) = \mathbf{Z}_p$, and $\pi^*(K_{Y_N})$ is nontrivial and thus generates $\pi^*(K_{Y_N}^{\perp})$. But if E is an exceptional curve on Y_N, then its class is of the form $e_9 + \check{\delta} + mK_{Y_N}$ so $\pi^*(e_9 + \check{\delta} + mK_{Y_N}) = \pi^*(e_9) + m\pi^*(K_{Y_N}) = \pi^*(e_9) + \check{\delta}^2\pi^*(K_{Y_N})/2$ and $(e_9 + \check{\delta} + mK_{Y_N}) \otimes \mathcal{O}_C(-f) = (\check{\delta}^2/2 + 1)\pi^*(K_{Y_N})$. And if we choose a smooth point $q \in C$ such that

$\mathcal{O}_C(q - f) \in \pi^*(K_{Y_N}^{\perp})$, then $\mathcal{O}_C(q - f) = j\pi^*(K_{Y_N})$ for some integer j.

Hence to show there is an irreducible exceptional curve $E \neq E_9$ on Y_N such that the restriction to C of $[E] - K_{Y_N}$ is the divisor class in $\mathrm{Pic}(C)$ of q, it is enough to show that there is a $\check{\delta}$ with $\check{\delta}^2/2 \equiv j \pmod{p}$. I.e., it is enough to show that the congruence class $j \pmod{p}$ lies in the image of the map $\Delta \to \mathbf{Z}_p$ from the truncated cone $\Delta = \{a_0 \check{r}_0 + \cdots + a_7 \check{r}_7 \neq 0 : a_i \geq 0 \text{ for all } i, \text{ and } 3a_0 + 2a_1 + 4a_2 + 6a_3 + 5a_4 + 4a_5 + 3a_6 + 2a_7 \leq p\}$ to $\mathbf{Z}_p$ given by $\check{\delta} \to \check{\delta}^2/2 \pmod{p}$.

We now show that the map $\Delta \to \mathbf{Z}_p$ is surjective for all primes $p > 2$. (It is irrelevant but true and easy to check that the map is surjective for $p = 2$ too.) From the matrix $(**)$ above we see $\check{r}_1^2 = -4$, $\check{r}_7^2 = -2$ and $\check{r}_1 \cdot \check{r}_7 = -2$ which for $\check{\delta} = (x\check{r}_1 + y\check{r}_7)$ gives $\check{\delta}^2/2 = (x\check{r}_1 + y\check{r}_7)^2/2 = -(2x^2 + 2xy + y^2) = -[(x+y)^2 + x^2]$. But every element t of $\mathbf{Z}_p$ is a sum of two squares [HP, p.8] so we can always find integers $0 \leq a \leq b \leq p/2$ such that $-(a^2 + b^2)$ is congruent to t. Setting $a = x$ and $b - a = y$ we have $a^2 + b^2 = (x+y)^2 + x^2$. As long as a and b are not both 0, the element $x\check{r}_1 + y\check{r}_7$ is in Δ and hence t is in the image of the map $\Delta \to \mathbf{Z}_p$ for $t \neq 0$. So say $t = 0$. If $p \equiv 1 \pmod{8}$ or if $p \equiv -3 \pmod{8}$, then -1 is a quadratic residue modulo p so $a^2 + b^2 \equiv t \pmod{p}$ clearly has a solution with $a = 1$. And if $p \equiv 3 \pmod{8}$, then -2 is a quadratic residue modulo p. Thus $(z\check{r}_7 + \check{r}_5)^2/2 = -(z+2)^2 - 2 \equiv t \pmod{p}$ has a solution z, which we may assume satisfies $0 \leq (z+2) \leq (p-1)/2$. This is equivalent to having either $0 \leq z \leq (p-5)/2$ or $z = -1$ or $z = -2$. For $z\check{r}_7 + \check{r}_5 \in \Delta$ to hold, we merely need $0 \leq z$ and $2z + 4 \leq p$, hence $0 \leq z \leq (p-4)/2$. But a direct substitution shows $z = -2$ is not a solution, and $z = -1$ is a solution only for $p = 3$, in which case $\check{\delta} = \check{r}_6$ gives an element of Δ mapping to $t = 0$. Thus $t = 0$ is in the image of $\Delta \to \mathbf{Z}_p$ when $p \equiv 3 \pmod{8}$.

Finally we must show that $t = 0$ is in the image of $\Delta \to \mathbf{Z}_p$ if $p \equiv -1 \pmod{8}$. Let $\check{\delta} = x\check{r}_1 + 2\check{r}_2 + y\check{r}_7$; for $\check{\delta}$ to be in Δ we need $x, y \geq 0$ and $x + y + 4 \leq p/2$. But $-\check{\delta}^2/2 = (4x^2 + 56 + 2y^2 + 28x + 4xy + 16y)/2 = (x+y+4)^2 + (x+3)^2 + 3$, so as above we can find integers $0 \leq a \leq b \leq (p-1)/2$ with $a^2 + b^2 \equiv -3 \pmod{p}$. Thus taking $x = a - 3$ and $y = b - a - 1$ we are done unless the solution involves either $a = 0, 1, 2$ or $a = b$.

If $a = 1$, then $x = -2$ and $(x + y + 4)^2 + (x + 3)^2 + 3 \equiv 0 \pmod{p}$ becomes $(y+2)^2 + 4 \equiv 0 \pmod{p}$, and, since 4 is a square, we see -1 must be a square modulo p, which cannot occur if $p \equiv -1 \pmod{8}$. If either $a = 0$ or $a = b$, then -3 must be a quadratic residue: $a = 0$ implies $(y+1)^2 + 3 \equiv 0 \pmod{p}$, while $a = b$ implies $2a^2 + 3 \equiv 0 \pmod{p}$ (noting that $p \equiv -1 \pmod{8}$ guarantees that 2 is a quadratic residue). Now if we let $\check{\delta} = x\check{r}_7 + \check{r}_2$, we find $-\check{\delta}^2/2 = (x + 2)^2 + 3$ which therefore is congruent to 0 (mod p) for some $0 \leq x + 2 \leq (p-1)/2$. For $\check{\delta}$ to be in Δ, we need $0 \le x$ and $0 \le 2x + 4 \le p$, the latter being equivalent to $0 \le x + 2 \le (p - 1)/2$. The only problems occur if either -1 or -2 is the solution to $(x + 2)^2 + 3 \equiv 0 \pmod{p}$. If $x = -1$, then $4 \equiv 0$ and so $p = 2$ (contrary to hypothesis) while if $x = -2$, then $p = 3$ contradicting $p \equiv -1 \pmod 8$.

Finally, if $a = 2$, then $(y + 3)^2 + 7 \equiv 0 \pmod{p}$, so -7 must be a quadratic residue. But we can now assume that -3 is not a quadratic residue and together with $p \equiv -1 \pmod 8$ this implies by quadratic reciprocity that $p \equiv -1 \pmod{24}$. This time we let $\check{\delta} = x \check{r}_7 + (\check{r}_6 + \check{r}_4)$, and we find $-\check{\delta}^2/2 = (x + 4)^2 + 7$ which therefore is congruent to $0 \pmod{p}$ for some $0 \le x + 4 \le (p-1)/2$. For $\check{\delta}$ to be in Δ, we need $0 \le x$ and $0 \le 2x + 8 \le p$, the latter being equivalent to $0 \le x + 4 \le (p-1)/2$. The only problems occur if either -1, -2, -3 or -4 is the solution to $(x + 4)^2 + 7 \equiv 0 \pmod{p}$. For the solution x to be -1, -2, -3 or -4, we must, respectively, have $p = 2$, 11, 2, and 7. But none of these is congruent to $-1 \pmod{24}$. $\quad\square$

REFERENCES

[BM] E. Bombieri and D. Mumford, *Enriques classification of surfaces in characteristic p*, II, Complex analysis and algebraic geometry, North-Holland, Amsterdam (1974), 23–42.

[H1] B. Harbourne, *Automorphisms of* K3-*like rational surfaces*, Proceedings of the 1985 American Mathematical Society Summer Research Institute on Algebraic Geometry, Bowdoin College, Proc. Sympos. Pure Math. 46, Part 2, Amer. Math. Soc., Providence (1987, pp. 17–28).

[H2] ——, *A rational surface with infinite automorphism group and no antipluricanonical curve*, Proc. Amer. Math. Soc. **99** (1987), 409–414.

[H3] ——, *Blowings-up of* $\mathbf{P}^2$ *and their blowings-down*, Duke J. Math. **52** (1985), 129–148.

[H4] ——, *Complete linear systems on rational surfaces*, Trans. Amer. Math. Soc. **289** (1985), 213–226.

[H5] ——, *Very ample divisors on rational surfaces*, Math. Ann. **272** (1985), 139–153.

[HL] B. Harbourne and W.E. Lang, *Multiple fibres on rational elliptic surfaces*, Trans. Amer. Math. Soc. **307** (1988), 205–223.

[HM] B. Harbourne and H.P. Miranda, *Exceptional curves on rational numerically elliptic surfaces*, J. of Algebra, vol. 128, no. 2, February 1, 1990, pp. 405–433.

[HP] Hughes and Piper, *Finite projective planes*, Graduate Texts in Mathematics, v. **6**, Springer-Verlag, New York-Heidelberg-Berlin, 1973, pp. xii and 291.

[K] K. Kodaira, *On compact analytic surfaces* II, Ann. of Math. **77** (1963), 563–626.

[N] V.V. Nikulin, *On a description of the automorphism groups of Enriques surfaces*, Soviet Math. Dokl.(1) **30** (1984), 282–285.

DEPARTMENT OF MATHEMATICS, UNIVERSITY OF NEBRASKA, LINCOLN, NE 68588

Contemporary Mathematics
Volume **116**, 1991

Small Resolutions of Gorenstein Threefold Singularities

SHELDON KATZ

Introduction

Let $Y \to X$ be a small resolution of the Gorenstein threefold singularity $p \in X$. Then by [**R**] X is cDV, i.e. X is locally given by the equation

$$f(x, y, z) + t g(x, y, z, t) = 0,$$

where $f = 0$ defines a rational double point surface singularity (hereafter referred to as a RDP). The question addressed here is the converse question of which cDV singularities admit small resolutions. A complete answer can be given in the cA_n and cD_n cases. In the cA_n case, we have the following result.

THEOREM 1.1. *Suppose* $p \in X$ *is an isolated* cA_n *singularity (i.e. the general* $t \in m_p$ *cuts out an* A_n *surface singularity), and* $Y \to X$ *is a small resolution. Then the exceptional curve in* Y *is a chain of* n $\mathbf{P}^1$ *'s meeting transversely, and* X *has the form* $xy + g(z, t)$, *where* $g(z, t)$ *has* $n + 1$ *distinct branches at the origin. Conversely, any* X *as above admits a small resolution.*

Several others have previously considered related situations in the cA_n case including work done by R. Friedman, R. Miranda, D. Morrison, H. Pinkham, and N. Shepherd-Barron (much of these unpublished). In particular, related results appear in [**F**, p. 676] and [**SB**, Propositions 11 and 12]. Friedman proves the converse part of Theorem 1.1. Shepherd-Barron proves that an isolated cA_n singularity X with equation $xy + g(z, t) = 0$ admits a nontrivial small morphism $Z \to X$ (Z is not necessarily smooth) if and only if g is reducible.

The cD_n case is not as nice, primarily for two reasons: there are now several possibilities for the exceptional curve, and there are no good normal

---

1980 *Mathematics Subject Classification* (1985 *Revision*). Primary 14B05; Secondary 14B07.
This paper is in final form and no version of it will be submitted for publication elsewhere.

forms known for cD_n singularities (but see [**Ma**]). In this paper, we avoid the problem of proliferation of cases by focusing on the cD_4 case, which contains all of the ideas needed in the general case. No attempt is made here to give normal forms for cD_n singularities which admit a small resolution. Instead, given a cD_n equation, one must bring it to the form considered here by an analytic change of variables, and then apply the results of §2 to the transformed equation to see if the singularity admits a small resolution. See Example 3.3.

The motivation for this work comes from [**P2**, §8]. Pinkham observes that one can construct small resolutions as deformations of partial resolutions of Du Val singularities. But there is some ambiguity here; the same singularity can be constructed in many ways, by starting with any Du Val hypersurface section as the special fiber of the deformation. So this procedure does not give a classification, as observed in [**P2**, p. 367]. The first step in actually classifying small resolutions is to see which partial resolutions arise as a *generic* hypersurface section. That is what is done here in the cA_n and cD_4 cases. Restrictions on the exceptional configurations in the general case have been found by D. Morrison [**M**]. The methods described here have been generalized in recent joint work of D. Morrison and the author to the cD_n and cE_n cases, and will appear elsewhere.

The cA_n case is treated in §1. The cD_n case is treated in §2. The main results of that section are Theorems 2.1 and 2.3. Examples are given in §3.

The author thanks Dave Morrison and Henry Pinkham for helpful discussions during the preparation of this manuscript, and extends his warm thanks to the organizers of the Sundance conference, who provided a stimulating environment for discussing mathematics.

0. Notation and conventions

Let $p \in X$ be an isolated Gorenstein threefold singularity, and $\pi : Y \to X$ a small resolution, so that $C = \pi^{-1}(p)$ is a transverse union of smooth rational curves C_i [**P2**]. Also [**R**] X is cDV, i.e., X is locally given by the equation $f(x, y, z) + tg(x, y, z, t) = 0$, where $t \in m_p$ defines a rational double point surface singularity S of type K_n, where $K = A, D,$ or E. Furthermore $\overline{S} = \pi^{-1}(S)$ is a normal surface with at worst rational double points. If the RDP is of type K_n, and $t \in m_p$ is general, the singularity $p \in X$ will be referred to as a cK_n singularity.

As in [**M**], this partial resolution $\pi|_{\overline{S}} : \overline{S} \to S$ can be represented by a graph. Let $\lambda : \hat{S} \to \overline{S}$ be the minimal resolution of $\overline{S}$. Let Γ be the dual graph of the configuration of exceptional curves of $\pi|_{\overline{S}} \circ \lambda$, which is the Dynkin diagram of type K_n. Vertices corresponding to curves blown down by λ are denoted by solid circles • and vertices corresponding to curves blown down by $\pi|_{\overline{S}}$ are denoted by open circles ∘. This graph Γ, together with its solid and open circles will be called the dual graph of the partial resolution of the RDP.

The versal deformation spaces of the surfaces $S, \overline{S}$ and $\tilde{S}$ are used in the sequel. The versal deformation spaces $\mathscr{B}$ and $\tilde{\mathscr{B}}$ of the surfaces S and $\tilde{S}$, together with their versal families $\mathscr{S}$ and $\tilde{\mathscr{S}}$, are worked out in [T]. As for $\overline{S}$, let $\Gamma_1 \subset \Gamma$ be the subgraph consisting of the solid vertices of Γ and the edges of Γ which connect them. As described in [P1], the Weyl groups $W(\Gamma)$ and $W(\Gamma_1)$ act on $\tilde{\mathscr{B}}$. The quotient by the $W(\Gamma)$ action is $\mathscr{B}$; the quotient by $W(\Gamma_1)$ defines the versal deformation space $\overline{\mathscr{B}}$ of $\overline{S}$. The versal family $\tilde{\mathscr{S}}$ is a partial simultaneous resolution of $\mathscr{S}$, given by appropriately blowing up the pullback of $\mathscr{S}$ to $\overline{\mathscr{B}}$. $\overline{\mathscr{S}}$ will be described in more detail in the next section. These three families fit together into the diagram

$$
\begin{array}{ccccc}
\mathscr{S} & \leftarrow & \overline{\mathscr{S}} & \leftarrow & \tilde{\mathscr{S}} \\
\downarrow & & \downarrow & & \downarrow \\
\mathscr{B} & \leftarrow & \overline{\mathscr{B}} & \leftarrow & \tilde{\mathscr{B}}.
\end{array}
$$

As in [P2, §8], the strategy is to represent a singularity by a map $\nu : D \to \mathscr{B}$, where D is the germ of a smooth curve. A small resolution can be represented by lifting ν to $\overline{\nu} : D \to \overline{\mathscr{B}}$. The techniques introduced here allow one to find the lifting criterion explicitly in the cA_n and cD_n cases in terms of the equation for X, to analyze when the Y given by $\overline{\nu}$ is smooth, and finally, to check when the singularity of X given by ν is of the desired type, and not actually simpler (see [P2, p. 367]).

1. The cA_n case

THEOREM 1.1. *Suppose $p \in X$ is an isolated cA_n singularity and $Y \to X$ is a small resolution. Then the dual graph of the partial resolution of the RDP is*

$$
\overbrace{\circ - \circ - \cdots - \circ - \circ}^{n \ times}
$$

and X has the form $xy + g(z, t)$, where $g(z, t)$ has $n+1$ distinct branches at the origin. Conversely, any X as above admits a small resolution.

PROOF. The strategy of the proof is to analyze the versal deformations of the various partial resolutions of an A_n singularity. Let the notation be as above. Near C, the threefold Y will be induced from the versal family $\overline{\mathscr{S}}$ by a morphism $\overline{\nu} : D \to \overline{\mathscr{B}}$, where D is a smooth curve. This will be expressed as $Y = \overline{\nu}^{-1}(\overline{\mathscr{S}})$. The threefold $\overline{\nu}^{-1}(\overline{\mathscr{S}})$ must therefore be smooth. It turns out that this forces X to be a cA_m singularity for some $m < n$ unless Γ_1 is empty.

As in [T], in the case of an A_n singularity, $\mathscr{B}$ is the affine space $k^n(t_1, \ldots, t_n)$, and $\mathscr{S}$ is the relative hypersurface H with equation

$$
xy + z^{n+1} + t_1 z^{n-1} + \cdots + t_n = 0.
$$

$\tilde{\mathscr{B}}$ is the hyperplane $a_1 + \cdots + a_{n+1} = 0$ in the affine space $k^{n+1}(a_1, \ldots, a_{n+1})$, where the map $\tilde{\mathscr{B}} \to \mathscr{B}$ is given by the elementary symmetric functions

$$t_i = \sigma_{i+1}(\mathbf{a}), \qquad i = 1, \ldots, n.$$

Furthermore, $\tilde{\mathscr{S}}$ is the variety obtained as follows: take the relative hypersurface $\tilde{H}$ obtained by pulling back H to $\tilde{\mathscr{B}}$, which has the equation

$$xy + \prod_{i=1}^{n+1}(z + a_i) = 0.$$

Blow it up by taking the closure of the graph of the mapping $\tilde{\mu} : \tilde{H} \to \mathbf{P}^1_{(1)} \times \cdots \times \mathbf{P}^1_{(n)}$ given by

$$(z_{i0} : z_{i1}) = (x : \prod_{j=1}^{i}(z + a_j)), \qquad i = 1, \ldots, n.$$

This is $\tilde{\mathscr{S}}$.

It remains to describe $\overline{\mathscr{B}}, \overline{\mathscr{S}}$. Number the vertices of Γ consecutively, and consider $\Gamma_1 \subset \Gamma$. Γ_1 consists of a union of k maximal connected subgraphs of lengths j_i, $i = 1, \ldots, k$ which begin with vertex number b_i. Note that by definition, $b_i + j_i < b_{i+1}$. Then $W(\Gamma_1)$ is the product of the symmetric groups of order $j_i + 1$ on the disjoint sets of variables $a_{b_i}, \ldots, a_{b_i+j_i}$ for each i. Replacing the set of variables $a_{b_i}, \ldots, a_{b_i+j_i}$ by the set $\sigma_{1,i}, \ldots, \sigma_{j_i+1,i}$ of their first $j_i + 1$ elementary symmetric functions give coordinates for the affine space $\overline{\mathscr{B}}$ after restricting to the hyperplane in these coordinates corresponding to $a_1 + \cdots + a_{n+1} = 0$.

EXAMPLE. $\bullet - \circ - \bullet - \bullet - \circ$. Here $\sigma_{\alpha,1} = \sigma_\alpha(a_1, a_2)$, and $\sigma_{\alpha,2} = \sigma_\alpha(a_3, a_4, a_5)$. Then $\overline{\mathscr{B}}$ is the hyperplane $\sigma_{1,1} + \sigma_{1,2} + a_6 = 0$ in $k^6(\sigma_{1,1}, \sigma_{2,1}, \sigma_{1,2}, \sigma_{2,2}, \sigma_{3,2}, a_6)$.

To simplify the exposition, $\overline{\mathscr{S}}$ will be described only for the above example. The reader will immediately see how to proceed in the general case. Pulling back $\mathscr{S}$ to $\overline{\mathscr{B}}$ gives the relative hypersurface $\overline{H}$ with equation

$$xy + (z^2 + \sigma_{1,1}z + \sigma_{2,1})(z^3 + \sigma_{1,2}z^2 + \sigma_{2,2}z + \sigma_{3,2})(z + a_6) = 0.$$

$\overline{\mathscr{S}}$ is the resolution of the above given by taking the closure of the graph of the mapping $\overline{\mu} : \overline{H} \to \mathbf{P}^1 \times \mathbf{P}^1$ given by

$$\overline{\mu}(x, y, z, \sigma_{\alpha,\beta}, a_6) = ((x : z^2 + \sigma_{1,1}z + \sigma_{2,1}),$$
$$(x : (z^2 + \sigma_{1,1}z + \sigma_{2,1})(z^3 + \sigma_{1,2}z^2 + \sigma_{2,2}z + \sigma_{3,2}))).$$

Returning to the general case, suppose given a morphism $\overline{\nu} : D \to \overline{\mathscr{B}}$ such that $Y = \overline{\nu}^{-1}(\overline{\mathscr{S}})$ is smooth above $0 \in \overline{\mathscr{B}}$. Note that the singular threefold X can be induced from $\mathscr{S}$ by the morphism $D \to \mathscr{B}$ given by composing

$\overline{\nu}$ with the natural map $\overline{\mathscr{B}} \to \mathscr{B}$. Denote by $l = n - \sum_{i=1}^{k} j_i$ the number of open vertices of Γ. The fiber over 0 contains as exceptional set the union of l rational curves $C_1, \ldots, C_l$ which contain A_{j_i} singularities on the curves whose corresponding vertices are adjacent to the component of Γ_1 associated to the singularity. It will be shown that these k singularities cannot all be smoothed by taking a family of hypersurface sections.

Choose a local parameter t on D near $0 \in D$, where $\overline{\nu}(0) = 0$. The coordinates on $\overline{\mathscr{B}}$ pull back to functions on D, and so may be viewed as functions of t. All of these functions vanish at 0 by definition.

Let m be the number of variables a_i which appear explicitly in the list of coordinates for $\overline{\mathscr{B}}$. Then the coefficient of z^m in the equation of $\overline{H}$ is $\prod_{i=1}^{k} \sigma_{j_i+1, i} + O(k+1)$, where $O(k+1)$ denotes terms of order at least $k+1$ at $t = 0$. In order for X to be a cA_n singularity (and not cA_m for some $m < n$), the coefficient of z^m must have order at least $n + 1 - m$ at $t = 0$. Since $k < n + 1 - m$, at least one of the $\sigma_{j_i+1, i}$ must vanish to order at least 2 at $t = 0$.

LEMMA 1.2. *Near the A_{j_i} singularity, Y is analytically isomorphic to the hypersurface with equation*

$$xy + z^{j_i+1} + \sigma_{1, i} z^{j_i} + \cdots + \sigma_{j_i+1, i} = 0,$$

where the $\sigma_{\alpha, i}$ are considered as functions of t.

PROOF. A straightforward computation. Suppose that the pullback of $\mathscr{S}$ to $\overline{\mathscr{B}}$ is the relative hypersurface $\overline{H}$ with equation of the form

$$xy + \prod_{\gamma=1}^{r} P_\gamma(z, \sigma_{\alpha, \beta}, a_\delta),$$

with a particular index $\gamma = \gamma_0$ corresponding to $P_{\gamma_0} = z^{j_i+1} + \sigma_{1, i} z^{j_i} + \cdots + \sigma_{j_i+1, i}$. Assuming for ease of notation that $1 < \gamma_0 < r$ (the remaining special cases are simpler), choose coordinates $(u_{\gamma_0-1} : v_{\gamma_0-1})$, $(u_{\gamma_0} : v_{\gamma_0})$ for $\mathbf{P}^1_{(\gamma_0-1)}$ and $\mathbf{P}^1_{(\gamma_0)}$. The A_{j_i} singularity occurs in the open subset given by $v_{\gamma_0-1} = v_{\gamma_0} = 1$. Substituting $x = u_{\gamma_0-1} P_1 \cdots P_{\gamma_0-1}$, $y = -u_{\gamma_0} P_{\gamma_0+1} \cdots P_r$ gives the local equation of $\overline{\mathscr{S}}$

$$u_{\gamma_0-1} u_{\gamma_0} = P_{\gamma_0},$$

which proves the lemma. Q.E.D.

Lemma 1.2 and the paragraph preceding it now imply that Y has a singular point if the dual graph contains a solid vertex. For in this case, the set of the $\sigma_{j_i+1, i}$ is nonempty, hence at least one of them vanishes to order at least 2 at $t = 0$, hence Y is singular at the corresponding A_{j_i} singularity of $\overline{S} \subset Y$. This proves the statement about the dual graph. It now follows that Y is locally induced from $\mathscr{S}$ by a map from a curve to $\overline{\mathscr{B}}$. This map gives an

explicit factorization of the branches of $g(z, t)$, which completes the proof
of the first part of Theorem 1.1. The converse is similar (and also shown in
[**F**, p. 676]). Q.E.D.

2. The cD_n case

Let $p \in X$ be an isolated cD_n singularity. As in [**T**], $\mathscr{S} \to \mathscr{B}$ is given
by

$$\mathscr{S} = \{(x, y, z, \mathbf{t}) \mid f(x, y, z, \mathbf{t}) = x^2 + y^2 z - z^{n-1} \\ - t_1 z^{n-2} - \cdots - t_{n-1} + 2t_n y = 0\},$$

and

$$\mathscr{B} = k^n(\mathbf{t}) = k^n(t_1, \ldots, t_n).$$

X can be given (but not uniquely) by a morphism $\nu : D \to \mathscr{B}$, where D
is the germ of a smooth curve. Let t be a local parameter on D. In other
words, we view the t_i as analytic fuctions of t, vanishing at $t = 0$.

We associate to f the function

$$F(z, t) = z^n + t_1 z^{n-1} + \cdots + t_{n-1} z + t_n^2.$$

F is a polynomial in z, and is analytic in t. The behavior of the branches
of the analytic plane curve $F(z, t) = 0$ determines whether or not X has a
small resolution.

THEOREM 2.1. *Let $p \in X$ be a cD_n singularity given by $\nu : D \to \mathscr{B}$ as
above. Then X admits a small resolution with dual graph*

*if and only if $F(z, t)$ factors into n distinct factors, each tangent to $z = 0$
with even mutiplicity.*

PROOF. By [**T**], $\mathscr{S} \times_{\mathscr{B}} \tilde{\mathscr{B}}$ is given by

$$x^2 + y^2 z - \frac{\prod_{i=1}^n (z + a_i^2) - a_1^2 \cdots a_n^2}{z} + 2a_1 \cdots a_n y = 0,$$

where $\tilde{\mathscr{B}} = k(a_1, \ldots, a_n)$. In other words, $F(z, t) = \prod_{i=1}^n (z + a_i^2)$. A map
$\tilde{\nu} : D \to \tilde{\mathscr{B}}$ lifting ν would give the a_i as analytic functions of t, yielding
the required factorization, and the converse is just as easy.   Q.E.D.

We now specialize to the cD_4 case.

DEFINITION 2.2. *Let C be a germ of an analytic branch of a plane curve.
Then C is an* odd hyperflex *if it is a nonordinary double point whose unique
tangent line meets it with odd multiplicity.*

THEOREM 2.3. *Let $p \in X$ be a cD_4 singularity given by $\nu : D \to \mathscr{B}$ as
above. If X admits a small resolution, then up to symmetries of the Dynkin*

diagram, the dual graph must be one of the following:

(i)

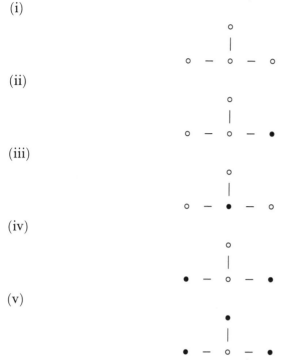

(ii)

(iii)

(iv)

(v)

Furthermore, these cases occur if and only if, respectively,

(i) $F(z, t)$ *factors into 4 distinct linear factors, each tangent to* $z = 0$
     with even multiplicity,

(ii) $F(z, t)$ *factors into 3 distinct factors, the first two of which are tangent
     to* $z = 0$ *with even multiplicity, and the third branch being an odd
     hyperflex, not tangent to* $z = 0$.

(iii) *Same as* (ii).

(iv) $F(z, t)$ *factors into 2 distinct factors, each an odd hyperflex, not
     tangent to* $z = 0$.

(v) $F(z, t)$ *factors into 3 distinct factors, the first two of which are smooth
     and transverse, the third branch being an odd hyperflex, not tangent
     to* $z = 0$.

PROOF. Case (i) is already done. For the remaining cases, first note that
$W(\Gamma)$ can be identified with the subgroup of $\mathrm{GL}(4)$ generated by permutations of the coordinates $a_1, \ldots, a_4$ and sign changes of the coordinates with
even product (follows from [T]). In case (ii), $W(\Gamma_1)$ is the subgroup generated by the permutation of the third and fourth coordinates. Hence $\overline{\mathcal{B}}$ has
coordinates $a_1, a_2, \sigma_1 = a_3 + a_4, \sigma_2 = a_3 a_4$. Making this substitution, we
see that

$$F = \prod_i (z + a_i^2) = (z + a_1^2)(z + a_2^2)((z - \sigma_2)^2 + \sigma_1^2 z).$$

Suppose that a small resolution $Y \to X$ is given by lifting ν to $\overline{\nu} : D \to \overline{\mathscr{B}}$.

LEMMA 2.4. *Near the A_1 singularity, Y is analytically isomorphic to the hypersurface with equation*

$$xy + z^2 + \sigma_1 z + \sigma_2 = 0.$$

PROOF. Analogous to the proof of Lemma 1.2. See [T, pp. 72–73] for some relevant calculations.   Q.E.D.

Since Y is smooth, Lemma 2.4 implies that σ_2 vanishes to first order at $t = 0$. It is immediate now that $(z - \sigma_2)^2 + \sigma_1^2 z = 0$ is an odd hyperflex, not tangent to $z = 0$. Conversely, an easy calculation shows that any odd hyperflex, not tangent to $z = 0$, can be put in this form after a change in variables in t alone. Since this change does not alter the line $z = 0$ of tangency to the first two branches, we may solve for all of the local coordinates of $\overline{\mathscr{B}}$ as functions of t, i.e., we can lift ν. This finishes case (ii). Case (iii) is essentially the same calculation, since $W(\Gamma_1)$ is generated by permutations of the second and third coordinates in this case. Case (iv) is similar. Finally, in case (v), $W(\Gamma_1)$ is generated by the three involutions $(a_1, \ldots, a_4) \to (a_2, a_1, a_3, a_4), (a_1, a_2, a_4, a_3), (a_1, a_2, -a_4, -a_3)$. Hence $\overline{\mathscr{B}}$ has coordinates $\sigma_1 = a_1 + a_2, \sigma_2 = a_1 a_2, s_2 = a_3^2 + a_4^2, \sigma_2' = a_3 a_4$, and

$$F = \prod_i (z + a_i^2) = ((z - \sigma_2)^2 + \sigma_1^2 z)(z^2 + s_2 z + \sigma_2'^2).$$

LEMMA 2.5. *In case (v), near the three A_1 singularities, Y is analytically isomorphic to*

$$xy + z^2 + \sigma_1 z + \sigma_2 = 0 \quad (\textit{first } A_1),$$

$$x^2 + y^2 z - z - s_2 + 2\sigma_2' y = 0 \quad (\textit{second and third } A_1).$$

PROOF. Another computation. Note that $x^2 + y^2 z - z$ is singular at the points $(0, \pm 1, 0)$. Lemma 2.5 implies that Y is smooth if and only if σ_2 and $s_2 \pm 2\sigma_2'$ each have order 1 at $t = 0$ (use coordinates centered at the singular points to see this). Since the discriminant of $z^2 + s_2 z + \sigma_2'^2$ is $s_2^2 - 4\sigma_2'^2 = (s_2 + 2\sigma_2')(s_2 - 2\sigma_2')$, the result follows easily.

Finally, as in the proof of Theorem 1.1, if the dual graph is distinct from these five types (modulo symmetries of the Dynkin diagram), then the singularity of X was in actuality simpler than cD_4. For example, if the dual graph were

then $\overline{\mathscr{B}}$ has local coordinates $\sigma_1 = a_1 + a_2 + a_3, \sigma_2 = a_1 a_2 + a_2 a_3 + a_3 a_1, \sigma_3 = a_1 a_2 a_3$, and a_4. As above, we see that σ_3 must have order 1 at $t = 0$. Computing the equation for X in terms of these coordinates, we see that the quadratic part of X has rank 2, hence X must be a cA_n singularity.   Q.E.D.

REMARKS. 1. It is possible to read off the sequence of normal bundles [M, P2] for each of the exceptional curves from the equation of X.

2. If X has a small resolution, it is well known that the resolution is not unique [P2]. Using the above analysis, we can see that in case (i), there are 192 ($=| W(\Gamma) |$) distinct small resolutions, in cases (ii) and (iii), there are 16 distinct small resolutions of each type, 8 in case (iv), and 2 in case (v). One merely counts the number of distinct ways of factoring $F(z, t)$ in terms of the desired local coordinates in each case (these numbers can also be calculated by [P2, Theorem 3]). Furthermore, the birational map between distinct small resolutions can be explicitly given in terms of blow-ups and blow-downs. Note that for a fixed X satisfying the hypotheses for cases (ii) and (iii), X admits small resolutions of both types (ii) and (iii).

3. Examples

According to [B, Satz 0.2], a base change of the smoothing of a rational double point admits a small resolution if and only if the order of the base change is a multiple of the Coxeter number of its associated Dynkin diagram. The first two examples illustrate this for A_n and D_n singularities.

EXAMPLE 3.1. Consider the base change $xy + z^{n+1} - t^k$ of the smooth threefold $xy + z^{n+1} - s$. By Theorem 1.1, this threefold admits a small resolution if and only if k is a multiple of $n + 1$ (the Coxeter number for A_n). Note that $z^{n+1} - t^{m(n+1)} = \prod_{j=0}^{n}(z - e^{2\pi i j/(n+1)}t^m)$.

EXAMPLE 3.2. Consider the base change $x^2 + y^2 z - z^{n-1} + t^k$ of the smooth threefold $x^2 + y^2 z - z^{n+1} + s$. Here, $F(z, t) = z^n - t^k z$. By Theorem 2.1, this threefold admits a small resolution with dual graph containing only open vertices if and only if k is a multiple of $2n - 2$ (the Coxeter number for D_n). Note that $z^n - t^{m(2n-2)}z = z \prod_{j=0}^{n-2}(z - e^{2\pi i j/(n-1)}t^{2m})$. One can also see by these methods that the same result is true without restriction on the dual graph, but this requires a bit more detail on the generalization of Theorem 2.3 to the cD_n case than is included here.

EXAMPLE 3.3. The first example of a cD_4 singularity with small resolution of type 2.3 (iv) was given by Laufer in [L, Example 2.3]. The example is the threefold X given by

$$u_1^2 + u_2^3 + u_3 u_4^2 + u_3^3 u_2 = 0.$$

To realize this singularity as a map from a curve to $\mathscr{S}$, first make the linear change of coordinates $u_1 = x$, $u_2 = t - z$, $u_3 = z$, $u_4 = y$ to get the new equation

$$x^2 + y^2 z - z^3 - z^4 + z^3 t + 3tz^2 - 3t^2 z + t^3 = 0.$$

This is not yet in the form of the versal deformation of a D_4 singularity described at the beginning of §2. To do this an *analytic* change of coordinates is needed (assume $k = C$, or else proceed formally). Such a coordinate change is most easily found by computer, using undetermined coefficients to find the first few terms of the series expansion. One such coordinate change is

$$y \mapsto (y + yz/6 + \cdots) + t(y/6 + yz/12 + \cdots)$$
$$+ t^2(11y/72 - yz/432 + \cdots) + t^3(-89y/1296 - 103yz/7776 + \cdots)$$
$$+ t^4(-2183y/31104 - 37yz/62208 + \cdots) + \cdots ,$$
$$z \mapsto (z - z^2/3 + \cdots) + t(-z/3 + \cdots) + t^2(-2z/9 + 5z^2/27 + \cdots)$$
$$+ t^3(22z/81 - 20z^2/243 + \cdots) + t^4(23z/243 - 10z^2/81 + \cdots) + \cdots .$$

In the above expressions, the $\cdots$ within parentheses denote omitted terms in y and z of order at least 3 (in particular, involving z alone in the second expression). The final $\cdots$ denote omitted expressions divisible by t^5. The equation of X is transformed to

$$x^2 + y^2 z - z^3 - z^2 t(-3 + t + t^2 + \cdots) - zt^2(3 - t - 2t^2/3 + \cdots) - (-t^3 + \cdots) = 0 ,$$

where all omitted terms have order at least 6. Using the notation of §2,

$$F(z, t) = z(z - t + t^2/3 + \cdots)((z - t)^2 + 2zt^2/3 + t^3/3 + \cdots) ;$$

hence X has a small resolution of type (v), by Theorem 2.3.

REFERENCES

[B] E. Brieskorn, *Die Auflösung der rationalen Singularitäten holomorphen Abbildungen*, Math. Ann. **178** (1968), 255–270.

[F] R. Friedman, *Simultaneous resolution of threefold double points*, Math. Ann. **274** (1986), 671–689.

[L] H. Laufer, *On* **CP**[1] *as an exceptional set*, Recent Developments in Several Complex Variables, Ann. of Math. Stud. **100** (1981), 261–276.

[Ma] D. Markushevich, *Canonical singularities of three-dimensional hypersurfaces*, Math. USSR-Izv. **26** (1986), 315–345.

[M] D. Morrison, *The birational geometry of surfaces with rational double points*, Math. Ann. **271** (1985), 415–438.

[P1] H. Pinkham, *Résolution simultanée de points doubles rationnels*, Séminaire sur les Singularités des Surfaces, Springer Lecture Notes in Math., vol. 777 Springer-Verlag, Berlin and New York, 1980, pp. 179–204.

[P2] H. Pinkham, *Factorization of birational maps in dimension 3*, Singularities, Proc. Sympos. Pure Math., vol. 40, Amer. Math. Soc., Providence, 1981, pp. 343-372.

[R] M. Reid, *Minimal models of canonical 3-folds*, Algebraic Varieties and Analytic Varieties Adv. Stud. Pure Math. **I** (1981), 131–180.

[SB] N. Shepherd-Barron, *The topology of rational 3-fold singularities*, Preprint, Cambridge.

[T] G. Tyurina, *Resolution of singularities of flat deformations of double rational points*, Functional Anal. Appl. **4** (1970), 68–73.

DEPARTMENT OF MATHEMATICS, OKLAHOMA STATE UNIVERSITY, STILLWATER, OK 74078

Contemporary Mathematics
Volume **116**, 1991

Multiple Tangents of Smooth Plane Curves (after Kaji)

STEVEN L. KLEIMAN

ABSTRACT. Hajime Kaji recently proved the following theorem about a complete, reduced, and irreducible plane curve C of degree at least 2 over an algebraically closed field of positive characteristic: *If C is smooth, or if C is nodal (that is, immersed) of geometric genus g at least 2, then C has only finitely many tangents making two or more distinct contacts.* In this paper, the significance of Kaji's theorem will be explained by placing it in a context of related results; then Kaji's theorem will be proved in detail.

1. Introduction

Hajime Kaji filled an important gap in our knowledge of plane projective geometry in characteristic p recently. He proved the following theorem about a complete, reduced, and irreducible plane curve C of degree at least 2 over an algebraically closed field.

THEOREM (Kaji, [14], [15]). *If C is smooth, or if C is nodal (that is, immersed) of geometric genus g at least 2, then C has only finitely many tangents making two or more distinct contacts. In other words, the Gauss (rational) map $\gamma\colon C \to C'$ is purely inseparable, where C' is the dual curve.*

Kaji's theorem is the subject of this paper: the theorem will be placed in context and proved in detail.

Kaji stated the main part of the theorem—the part about a nodal curve C with $g \geq 2$—for a curve X in $\mathbf{P}^N$ with $N \geq 2$. However, he began by observing that X may be replaced by a general projection C of itself into the plane. Thus, the additional generality is not significant.

Kaji proved that the theorem is sharp: for $g = 0, 1$, he constructed a nodal curve C whose Gauss map γ is not purely inseparable. The basic idea is simple. Given an (abstract) smooth curve X of genus g, find a

1980 *Mathematics Subject Classification* (1985 *Revision*). Primary 14N05; Secondary 14N10, 14H99.

Supported in part by NSF grant 8801743 DMS.

This paper is in final form and no version of it will be submitted for publication elsewhere.

smooth curve X' and an embedding ι of X into the product $Y' := X' \times \mathbf{P}^1$ such that the projection $X \to X'$ is finite and inseparable, but not purely inseparable. Then embed Y' into a $\mathbf{P}^N$ so that the fibers of Y'/X' are lines. Those lines are obviously the embedded tangent lines. So a general plane projection C of X is a nodal curve whose Gauss map γ is not purely inseparable. The problem is to find an X' and an $\iota: X \hookrightarrow Y'$ that work.

When $g = 0$, it is relatively easy to find an X' and an ι that work. In fact, Kaji had already done so in an earlier paper ([13], 4.1, p. 439). Independently, Jürgen Rathmann ([25], 2.13, p. 576) discovered the basic idea and treated the case $g = 0$.

When $g = 1$, the problem is more difficult, and Kaji devoted about a third of his paper [14] to it. (Kaji's paper [15] appears to be a revision of [14], with some additions and refinements.) He found a remarkable situation: suitable X' and ι exist if and only if X is an ordinary elliptic curve; in fact, the conclusion of the theorem holds if C is a nodal curve whose normalization X is a supersingular elliptic curve, and the conclusion fails for some nodal curve C whose normalization X is any given ordinary elliptic curve. Kaji's reasoning, although important and interesting, is somewhat special to the case $g = 1$, and the matter will not be discussed further in this paper.

In §§2 and 3, we shall place Kaji's theorem in context. In §2, we shall recall some results of Abramo Hefez, Anders Thorup, Andrew Wallace, and the author that imply this: on the one hand, if C is general among the curves of its degree and if $p \neq 2$, then the Gauss map γ is in fact birational; on the other hand, there exists a singular curve C such that γ has an arbitrarily large separable degree.

The construction of the latter curve C does not yield an explicit equation for it. So, in §2, we shall discuss an explicit example, which was worked out at the conference by Dan Laksov and the author. In the example, as it happens, C's tangents are concurrent (or C is a "strange" curve); in other words, C' is a line.

If the tangents of a plane curve C are concurrent and C is smooth or nodal (but not a line), then C is a conic and $p = 2$. That theorem was proved by Emilio Lluis ([23], p. 51) in 1961; in fact, Lluis considered a nodal curve in $\mathbf{P}^N$, but again we may reduce the case $N \geq 3$ to the case $N = 2$ by projecting the curve. The theorem was rediscovered for a smooth curve in $\mathbf{P}^N$ by Pierre Samuel ([27], p. 76) in 1966. In 1981, Dan Laksov ([22], p. 214) gave a simple, elegant proof. Laksov considered a smooth curve in $\mathbf{P}^N$, but his proof works equally well for a nodal curve, as Kaji ([15], Remark, p. 9) asserts. We shall review Laksov's proof in §2 and check the nodal case. We shall also mention some interesting related results of Valmecir Bayer, Hefez, Audun Holme, Lluis, and Israel Vainsencher; most of the results were not available until well after the conference.

Suppose $p = 2$. Then the Gauss map γ is inseparable for every curve C, smooth or singular. That statement was well known in the 1950s; one elementary proof was given by Samuel ([26], Theorem 1, p. 13-02), and another proof may be given using a local parametrization (Jean-Pierre Serre, private communication). (In fact, the Gauss map of any odd-dimensional variety is purely inseparable; that statement was proved by Nicholas Katz ([16], Proposition 3.3, p. 221, and §1.2, p. 214)). Nevertheless, if C is general among the curves of its degree, then γ is as good as possible: it is purely inseparable of degree 2. This theorem was proved at the conference by William Lang and the author. The proof proceeds by induction on the degree of C via a degeneration technique, and it will be presented in §3.

The latter theorem, it turned out, can be refined: if C is smooth or "moderately singular" and if $p = 2$, then γ is purely inseparable of degree 2 unless the degree of C is odd and at least 5 and unless the equation of C has a certain special form. That statement follows immediately from the inseparability of γ, Kaji's theorem, and a significant generalization of a lovely theorem of Rita Pardini [24]. The generalization was proved using Hasse–Schmidt differential operators independently by Hefez for a moderately singular C ([6], (5.10) p. 21, (5.16) p. 25—that work was available in preprint form in April 1987) and by Masaaki Homma for a smooth C ([12], Corollary 2.5). These matters will be discussed further in §3.

In §4, we shall prove Kaji's theorem in detail. We shall give a version of Kaji's proof, which is a little simpler than the original one and which clarifies the major new idea (it is also clarified in Kaji's second version [15]). The simplification comes from using intersection theory on the point–line incidence correspondence and from using one of the standard Plücker formulas. Kaji's new idea is to study the geometry of certain curves on Y', where Y' is the normalized ruled surface of tangents to C. The base curve of Y' is the normalization X' of C'. And, in the case of most interest, X' is not equal to the normalization X of C. Kaji's proof involves some delicate calculations with intersection numbers on Y', and until the end of the proof, it is unclear that it will pay off to do those calculations. The proof is characteristic free, but of no interest in characteristic zero.

2. Basic notions and examples

A plane curve C (not a line) is called *reflexive* if the following condition is satisfied: given a line l and a point P on it, then P is a general point of C, and l is the line tangent at P if and only if l corresponds to a general point of the dual curve C', and P corresponds to the line tangent there. In other words, C is equal to the dual C'' of C', and the Gauss maps γ of C and γ' of C' are inverse birational correspondences.

Gaspard Monge in 1805 gave a simple argument involving derivatives (see [20], p. 166) that works virtually without change in any characteristic p and shows that C is reflexive if (and only if) the Gauss map γ is separable. (Wallace [28] was the first to seriously consider the case where $p > 0$.) In particular, if $p = 0$, then C is always reflexive, whether it is smooth or not; so then C has only finitely many tangents making two or more distinct contacts.

The condition that $C = C''$ is called *reciprocity*. Reciprocity may hold but reflexivity fail, even if C is smooth. For example ([28], §7.2, p. 340; [20], p. 88), the curve,

$$(1) \qquad C : y = x^{q+1} - y^q \qquad \text{where } q := p^e \text{ for some } e \geq 1,$$

is smooth (also at infinity) and satisfies reciprocity, but it is not reflexive. In fact, the dual curve C' has, in appropriate coordinates, exactly the same equation as C, and the Gauss map γ is equal to the Frobenius qth power map Φ_q. Hefez ([6], (6.7.4) p. 26 and (4.11) p. 15) asked if C is the only example, up to projective equivalence, of a smooth nonreflexive curve satisfying reciprocity.

The answer is no. Indeed, Garcia and Voloch recently proved this lovely theorem ([5] Theorem 4, p. 17): a nonreflexive, smooth plane curve of degree at least 4 defined over the field with $q := p^e$ elements satisfies reciprocity if and only if it is "Frobenius nonclassical" in the sense that the tangent line at an arbitrary simple point P contains the image of P under the Frobenius qth power map Φ_q. Hefez and Voloch [9] made a general study of Frobenius nonclassical curves, and Garcia [4] gave examples of smooth ones other than the Fermat curve.

If a plane curve C is general (in particular, smooth) and if $p \neq 2$, then C is reflexive by a theorem of Hefez and the author ([7], (5.6), p. 176). In this case, γ is not only purely inseparable, it is birational. Of course, the curve of Example (1) is smooth, and its Gauss map γ is purely inseparable of degree q, where q can be taken arbitrarily large.

In 1958, Andreotti ([1], Lemma, p. 826) proved that, if C is the general plane projection of a nonhyperelliptic canonical curve, then its Gauss map γ is purely inseparable. He used this fact to prove Torelli's theorem in positive characteristic. Today, the fact may be viewed as an example of Kaji's theorem. Andreotti also showed ([1], p. 826) that C is determined by C'.

On the other hand, there exists a curve C whose dual C' is any given (reduced and irreducible) curve and whose Gauss map γ has any given separable degree s and inseparable degree $q = p^e$ provided only that $p > 0$ and $e \geq 1$. In particular, there are infinitely many C with the same dual C'. The existence of such C was discussed by Wallace ([28], §7.3, p. 340–341), and his discussion was refined by Thorup and the author ([20], pp. 170–171). It would be good to know when a smooth such C exists. By Kaji's theorem, a necessary condition is that $s = 1$. By Lluis's theorem, if C' is a line, a necessary and sufficient condition is that $q = 2$. But are these conditions the only restrictions? In other words, is every curve C' of degree at least two the dual of some smooth curve C whose Gauss map has any given inseparable degree $q = p^e$?

In the preceding paragraph, C' and q are arbitrary; so there exists a nonreflexive curve C whose dual curve C' is either reflexive or not, as desired. If C' is reflexive, then obviously C does not satisfy reciprocity. Now, in fact, there exists a *smooth* nonreflexive C such that C' is reflexive. Indeed, Pardini ([24], §5) gave the following example:

$$C : x^{rp} + xy^{rp} + y = 0, \qquad r > 1, \ r \not\equiv 0 \, (p).$$

That C is smooth and its Gauss map γ is inseparable of degree p. Furthermore, C' is reflexive if and only $r \not\equiv 1 \, (p)$. Independently, Homma (private communication, see [21], p. 342) gave another explicit example of a smooth curve not satisfying reciprocity.

Wallace's construction also shows that there exists a curve C such that a general tangent makes an arbitrarily large number of distinct contacts, that number being the separable degree s of γ. However, the construction does not yield an explicit equation for C. So here is a particular example, found by Laksov and the author:

$$(2) \qquad C : xy^{q-1} + x^q + y^{q^2} = 0 \qquad \text{where } q := p^e > 0.$$

To analyze C, differentiate the equation implicitly with respect to x, getting

$$y^{q-1} - xy^{q-2}y' = 0 \qquad \text{or} \qquad y' = y/x.$$

Therefore, every tangent line passes through the origin. Now, C is clearly invariant under the transformation

$$(x, y) \mapsto (\zeta x, \zeta y) \qquad \text{where} \quad \zeta^{q-1} = 1.$$

Moreover, every line through the origin is invariant under that transformation. Therefore, every line that makes a contact with C outside the origin makes at least $q - 1$ distinct contacts. Since $q = p^e$, we can take $q - 1$ to be arbitrarily large.

To prove that C is irreducible, set $z := x/y$ and $t := y^q$. Then its equation becomes

$$z + z^q + t^{q-1} = 0.$$

Replace t by $t+z$. Then it suffices to prove that for $r := q-1$ the polynomial

$$(z + t)^r + z^q + z = t^r + rt^{r-1}z + \cdots + rt^{r-1} + z^r + z^q + z$$

is irreducible. If $r = 1$, the polynomial is trivially irreducible. If $r \geq 2$, it is irreducible by Eisenstein's criterion, applied over the polynomial ring $k[z]$ where k is the ground field.

It is easy to check that the preceding curve has a cusp (at infinity), in accordance with the following theorem, which asserts that the only example of smooth or nodal curve (irreducible and not a line) whose tangents are concurrent is the conic in characteristic 2.

THEOREM (Lluis's theorem ([23], p. 51)). *If a plane curve* C *of degree* d *at least* 2 *is smooth or nodal and if all its tangents contain a common point* P, *then* C *is a conic and* $p = 2$. *Moreover, the converse holds.*

Assuming C smooth, Laksov ([22], p. 214) gave a simple, coordinate-free proof. That proof works equally well when C is nodal, as Kaji ([15], Remark, p. 9) remarked and we shall now see.

Consider the following standard exact sequence of locally free sheaves on the normalization X of C:

$$0 \longrightarrow \Omega_X^1(1) \longrightarrow \mathcal{P}_X^1(1) \longrightarrow \mathcal{O}_X(1) \longrightarrow 0$$

where the term in the middle is the sheaf of principal parts. Suppose C is immersed. Then $Y := \mathbf{P}(\mathcal{P}_X^1(1))$ is the total space of the family parametrized by X of tangent lines l to C (see for example, [18], IV, A, pp. 341–346). Suppose each l contains a common point P. Then the family of inclusion maps corresponds to a surjection,

$$\mathcal{P}_X^1(1) \longrightarrow\!\!\!\!\!\rightarrow \mathcal{O}_X.$$

That surjection cannot factor through $\mathcal{O}_X(1)$, because $\mathcal{O}_X(1)$ and $\mathcal{O}_X$ are not isomorphic as X is complete. Hence, the induced map,

$$\Omega_X^1(1) \longrightarrow \mathcal{O}_X,$$

is nonzero. Therefore, we have

$$0 \geq \deg \Omega_X^1(1) = 2g - 2 + d$$

where g is the genus of X and d is the degree of C. Consequently, $g = 0$ and $d = 2$ as $d \geq 2$. Finally, a simple computation shows that if C is smooth conic, then its tangents contain a common point P if and only if $p = 2$. Thus the theorem is proved.

A number of other interesting results have been proved about a plane curve C of degree d at least 2 whose tangents contain a common point P. First, Holme and Lluis [10] considered the projection from P. It is equal to the Gauss map γ; indeed, it has the right inseparable degree because of formula (7) in the next section. So

(3) $$d = n + m$$

where n is the degree of γ and m is the multiplicity of P on C (if $P \notin C$, then $m = 0$). Equation (3) yields this: $n = d$ if and only if $P \notin C$; and $n = d - 1$ if and only if P is a simple point of C. Thus we recover the following results of Lluis ([23], bottom of p. 47): if p does not divide d, then $P \in C$; if also p does not divide $d - 1$, then C is singular at P. In fact, Holme and Lluis worked with varieties of arbitrary dimension in $\mathbf{P}^N$ and proved related results; Bayer and Hefez ([2], Theorem 1) proved Equation (3) for a curve in $\mathbf{P}^N$.

Second, Bayer and Hefez ([2], §3, Corollary 1) proved that if s and q are the separable and inseparable degrees of γ (so $sq = n$), then C is projectively equivalent to a curve whose equation is of the form

$$(4) \qquad a_m(x)y^{sq} + a_{m+q}(x)y^{(s-1)q} + \cdots + a_{m+sq}(x) = 0$$

where a_i is a polynomial of degree i and $a_{m+jq} \neq 0$ for some j such that $s - j \not\equiv 0\,(p)$; moreover, the converse holds. Suppose that C is general for fixed d and q. Then ([2], end of §3) clearly s and m are equal to the quotient and remainder resulting from the division of d by q, and if $m > 1$, then P is an ordinary multiple point. Bayer and Hefez ([2], Proposition 4 and Theorem 6) went on to give a formula for the number of singular points of C other than P and to show that those points have distinct tangents; moreover, if $p \neq 2$, then the points are all cusps with semigroup $[2, q]$.

For instance, Example (2) fits into this context, an appropriate projective transformation being the one that interchanges the y-axis and the line at infinity. However, Example (2) is not general, because it has a single higher cusp; moreover, for it, $d = q^2$, $s = q - 1$, and $m = q$.

Third, Hefez and Vainsencher [8] determined the degree $b = b(d, q)$ of the variety of all curves C of degree d, with concurrent tangents, and whose Gauss map has inseparable degree q. Moreover, they show that b is the number of distinct C passing through the appropriate number of points in general position.

3. Moderately singular curves and extremal curves

While Wallace's construction does not yield an explicit (irreducible) equation for the desired curve C, it does show that C must satisfy a polynomial F of the form,

$$(5) \qquad F = D(x^q, y^q)y - M(x^q, y^q)x - B(x^q, y^q),$$

where D, M, and B are relatively prime polynomials and q is the inseparable degree of γ. The irreducible polynomial of C must divide F, but it need not be of the same form. For example (Hefez and Rathmann respectively, private communication, see [21], p. 341), if $p = 2$, then the irreducible polynomial of a smooth conic is not of that form, and if $p \geq 3$, then there are many nodal curves whose irreducible polynomials are not of that form.

Consider now a smooth and nonreflexive curve $C: F = 0$ of degree d at least 3. Pardini ([24], Corollary 2.2, p. 8, Proposition 3.7, p. 13) proved this: if $p \geq 3$, then F is of the form (5), but with q replaced by p; moreover, if $d = p + 1$, then C is projectively equivalent to the curve of Example (1).

Hefez and Homma significantly improved Pardini's theorem as follows. First Homma ([11], Theorem 6.1, p. 1490, Proposition 6.7, p. 1497) proved the equivalence of the following conditions: (a) $d = q + 1$; (b) C' is smooth; and (c) if $d \geq 4$, then C is projectively equivalent to the curve of Example (1);

if $d = 3$, *then* $p = 2$. In the latter case, by a theorem of Max Deuring ([3], p. 47), C is projectively equivalent to a curve whose defining polynomial F depends on the j-invariant; namely,

$$F = \begin{cases} x^3 + y^2 + y & \text{if } j = 0, \\ jx^3 + xy^2 + xy + 1/j & \text{if } j \neq 0. \end{cases}$$

Then Hefez ([6], (5.10) p. 21, (5.16) p. 23) and Homma ([12], Corollary 2.5) independently generalized the main part of Pardini's theorem, proving this: *if the inseparable degree q of the Gauss map γ of C is at least* 3, *then F is of the form* (5). In fact, Hefez allowed C to be *moderately singular* in the following sense:

$$\sum_{P \in C} e_P < d/2$$

where e_P is the multiplicity of the Jacobian ideal $(F_x, F_y)\mathcal{O}_{C,P}$ or, in terms of intersection multiplicities,

$$e_P = \min\left(i(P, \{F_x = 0\} \cdot C), \; i(P, \{F_y = 0\} \cdot C) \right).$$

(Hefez and Voloch ([9], §3) use different terminology: they say that C has *controlled singularities*.) A nodal curve is not necessarily moderately singular; for example, a nodal cubic is not.

Both Hefez and Homma went on to give similar simple proofs that (a), (b), and (c) are equivalent. Namely, assume (a). If $d = 3$, then trivially $p = 2$. Assume $d \geq 4$. Then, by the preceding result, F is of the form (5). And, an argument like Pardini's proof of Proposition 3.7 in [24] shows that, given (a) and $d \geq 4$, the polynomials of the form (5) may be transformed into one another by a linear change of variables. Thus (c) holds.

To proceed with the proof, recall the following standard Plücker formula (see for example [18], (I,24), p. 307, also pp. 309-310):

(6) $d(d-1) = nd'$ where $n := \deg(\gamma)$ and $d' := \deg(C')$.

By Kaji's theorem, γ is purely inseparable; hence, $n = q$, and the geometric genera g, g' of C, C' are equal. Of course, Hefez did not appeal to Kaji's theorem, which had not yet been proved, but he proved it in the case at hand.

Assume (c). If $d \geq 4$, then (b) holds because the curve of Example (1) and its dual share the same equation. If $d = 3$ and $p = 2$, then $d' \leq 3$ by (6), and $g' = 1$ as $g = 1$; so C' is smooth. Thus (c) implies (b). Finally, assume (b). Since $g = g'$ and both C and C' are smooth, $d = d'$. Since $n = q$, therefore (6) yields (a). The proof is now complete.

Consider an arbitrary curve C, and as always, let s, q, and n denote the separable, inseparable, and total degrees of the Gauss map γ (so $n = sq$). Obviously, a general tangent l makes s distinct contacts. Suppose $q > 1$. Then, at each point of contact, P, the intersection multiplicity $i(P, l \cdot C)$ is equal to q; the multiplicity is not more than q even though P is a point where q contacts coalesce. This formula

(7) $i(P, l \cdot C) = q$

was proved by Hefez and the author ([7], (3.5), p. 156). The author gave a much shorter and more conceptual proof in ([20], p. 175). The formula was rediscovered by Homma ([11], Proposition 4.4, p. 1486).

Formula (7) and Bezout's theorem yield the following bound:

(8) $$n \leq d.$$

Hefez ([6], (7.16) p. 33) proved that if $n = d$, then C' is a line. If in addition C is immersed, then by Lluis's theorem, C is a conic and $p = 2$. In particular, if C is smooth—whence by Kaji's theorem $n = q$—then the two extremal cases, $n = d$ and $n = d - 1$, are completely described by that result and by the equivalence of (a), (b), and (c). Finally, Homma ([11], 3.4, p. 1482) proved that whenever $n = q$ (that is, γ is purely inseparable), then $n = d$ if and only if C is projectively equivalent to a curve whose equation is of the form,

$$y^q = x^{q-1} + a_2 x^{q-2} + \cdots + a_{q-2} x^2 + x.$$

Bayer and Hefez ([2], Corollary 2) recovered that result from their normal form (4), after noting that Equation (3) implies $m = 0$.

On the other hand, if $p \neq 2$, then C is reflexive if and only if $i(P, l \cdot C) = 2$, and if $p = 2$, then C is not reflexive ([26], Theorem 1, p. 13-02), ([28], 6., p. 341), ([1], Corollary, p. 824) ([16], Proposition 3.3 p. 221, and §1.2 p. 214). The latter result and Example (1) raise the question of what happens when $p = 2$ and C is suitably general. That question is answered by the following theorem.

THEOREM. *Assume that C is immersed and moderately singular and that $p = 2$. Then the Gauss map γ is purely inseparable of degree 2, and a general tangent makes a single and simple contact, unless the degree d of C is odd, $d \geq 5$, and the irreducible polynomial F of C is of the special form (2).*

Here is the proof. First, recall that γ is inseparable. So n:= $\deg(\gamma)$ is even, and it suffices to prove $n = 2$.

Suppose $n \neq 2$. By Kaji's theorem, $n = q$, where q is the inseparable degree of γ. So, $q \geq 4$. Hence by the generalization of Pardini's theorem (recalled above), the irreducible polynomial F of C is of the form (5). In particular, $d = kq + 1$ where k is the maximum of the degrees of the polynomials D, M, and B in (5). Thus the theorem is proved.

Here is a different argument, which only proves the following less precise result: $q = 2$ if C is suitably general. This argument was found by W. Lang and the author. In addition, Ragni Piene and the author had a useful discussion about some of the details. The argument does not involve Kaji's theorem nor the generalization of Pardini's theorem, and it may be of independent interest.

Let **C** be the total space of the family of all plane curves of degree d, and **T** its parameter projective space. Let **C*** be the conormal space of the family: it is the closure in $\mathbf{P}^2 \times \check{\mathbf{P}}^2 \times T$ of the set of all triples (P, l, t) such that P is a simple point of the curve represented by t and l is the curve's tangent line at P. Let $\mathbf{C}'$ be the image of $\mathbf{C}^*$ in $\check{\mathbf{P}}^2 \times T$, and $\Gamma \colon \mathbf{C}^* \to \mathbf{C}'$ the induced map.

Let $\mathbf{U}'$ be the dense open subset of $\mathbf{C}'$ where Γ is finite and $\Gamma_* \mathcal{O}_{\mathbf{C}^*}$ is locally free, say of rank n. The image of $\mathbf{U}'$ in T is dense and constructible, so it contains a dense open subset V. Replace V by a smaller dense open subset of T so that

(1) the formation of $\mathbf{C}^*$, $\mathbf{C}'$, and Γ commute with base change through V (see [19], (3.10), p. 186), and

(2) the curves C parametrized by V are smooth.

Clearly, the Gauss map of a C belonging to V is of degree n.

The proof proceeds by induction on d. If $d = 2$ or 3, then $n = 2$ because $n \le d$ by (8) and n is even.

Suppose $d \ge 4$. By induction, there exists, for $i = 1, 2$, a smooth curve C_i of degree $d_i \ge 2$ with $d_1 + d_2 = d$, whose Gauss map is of degree 2. Replace C_1 by a general translate so that C_1 and C_2 meet transversally ([17], 11, p. 296). By the Plücker formula (6),

$$d_i(d_i - 1) = 2d_i' \qquad \text{where } d_i' := \deg(C_i').$$

Let S be a line in T meeting V and containing the point, s say, that represents $C_1 + C_2$. Let $\mathbf{C}_S$ be the restricted total space. Let $\mathbf{C}_S^*$ be the conormal scheme of $\mathbf{C}_S/S$. Let $\mathbf{C}_S'$ its image in $\check{\mathbf{P}}^2 \times S$, and $\Gamma_S \colon \mathbf{C}_S^* \to \mathbf{C}_S'$ the induced map. Then, clearly, Γ_S is of degree n. Moreover, $\mathbf{C}_S'$ is flat over S, because $\mathbf{C}_S'$ is reduced and S is a smooth curve.

The reduced fiber $\mathbf{C}_S^*(s)_{\mathrm{red}}$ consists of the conormal schemes C_1^*, C_2^* plus the conormal schemes P^* of the points P in $C_1 \cap C_2$. Each P^* is simply the fiber of the incidence correspondence I over P, and it appears for the following reason. The Gauss map of a plane curve is given by the partial derivatives of its defining polynomial. Since the conormal scheme $\mathbf{C}_S^*$ is the closure of the graph of the family of Gauss maps, therefore $\mathbf{C}_S^*$ is equal to the blow-up of $\mathbf{C}_S$ along the relative singular locus. In particular, each P^* supports a component of the exceptional divisor.

The reduced fiber $\mathbf{C}_S'(s)_{\mathrm{red}}$ consists therefore of the dual curves C_1', C_2' plus the lines P' dual to the points P in $C_1 \cap C_2$. Hence

$$\deg(\mathbf{C}_S'(s)_{\mathrm{red}}) = d_1' + d_2' + d_1 d_2.$$

Let $\eta \in S \cap V$. Then, because $\mathbf{C}_S'/S$ is flat,

$$\deg(\mathbf{C}_S'(\eta)) = \deg(\mathbf{C}_S'(s)) \ge \deg(\mathbf{C}_S'(s)_{\mathrm{red}}).$$

Moreover, by the Plücker formula (6),

$$n \deg(\mathbf{C}_S'(\eta)) = d(d - 1).$$

Therefore, since $d = d_1 + d_2$, the preceding displays yield

$$\tfrac{1}{n}(d_1^2 - d_1 + d_2^2 - d_2 + 2d_1 d_2) \geq \tfrac{1}{2}(d_1^2 - d_1) + \tfrac{1}{2}(d_2^2 - d_2) + d_1 d_2.$$

Since $n \geq 2$, therefore $n = 2$. The proof is now complete.

It is interesting to note a postiori that the fiber $C_S'(s)$ is reduced, because it has no embedded components and the inequalities above are in fact equalities. On the other hand, in the corresponding situation in arbitrary characteristic p, it can be shown that the conormal schemes P^* appear in the fiber $C_S^*(s)$ with multiplicity 2 and that, if $p \neq 2$, then each line P' appears with multiplicity 2.

Proof of Kaji's theorem

Kaji's theorem, which was stated at the beginning of the article, will now be proved in eleven small steps.

STEP 1: *The case of geometric genus g at most 1.*

In this case, C is smooth by hypothesis. Hence, C is of degree 2 or 3. Therefore, by Bezout's theorem, no tangent can make two or more distinct contacts.

STEP 2: *The setup.*

As before, let X and X' denote the normalizations of C and C'. Let g and g' denote the geometric genera, and assume that $g \geq 2$.

Let I be the graph of the point–line incidence correspondence, and form the following diagram with fiber squares:

$$
\begin{array}{ccccc}
Y & \xrightarrow{\eta} & Y' & \longrightarrow & I := \{P \in L\} \\
\downarrow & \square & \downarrow & \square & \downarrow \\
X & \longrightarrow & X' & \longrightarrow & \check{\mathbf{P}}^2
\end{array}
$$

where Y and Y' are defined as the indicated products. Then Y, resp. Y', is the total space of the family parametrized by X, resp. X', of tangent lines l to C.

Consider the (well-known) section $\sigma: X \to Y$ defined by placing a point of X on the associated tangent line l of C. Form the following curve on Y':

$$D := \eta\sigma(X).$$

Then D is birational to X. Indeed, the projection of I onto $\mathbf{P}^2$ obviously carries D onto C, and the composition $X \to D \to C$ is obviously equal to the normalization map. Hence, $X \to D$ is a birational map.

Let H denote the preimage in Y' of a general line in $\mathbf{P}^2$. Then H is irreducible; in fact, it is a section of Y'/X'. Indeed, let F be an arbitrary fiber of Y'/X'. Then the scheme-theoretic intersection $H \cap F$ is a reduced point, because the projection to $\mathbf{P}^2$ carries F isomorphically onto a tangent line to C. Therefore, the projection to X' carries H isomorphically onto X'. Thus, H is a section.

Finally, set

$$n := \deg(\gamma), \quad d := \deg(C), \quad \text{and} \quad d' := \deg(C').$$

STEP 3: $D \cdot F = n$.

Indeed, the intersection number $D \cdot F$ is equal to the degree $\deg(D/X')$, because both those numbers are equal to the length of the scheme-theoretic intersection $D \cap F$. Finally, $\deg(D/X') = n$ because $\eta\sigma: X \to D$ is birational.

STEP 4: $D \cdot H = d$.

Indeed, the projection $D \to C$ is birational, and H is the preimage of a general line.

STEP 5: $H^2 = d'$.

Indeed, let L, resp. L', be the preimage in I of a line in $\mathbf{P}^2$, resp. $\check{\mathbf{P}}^2$. Then the rational equivalence class of Y' on I is equal to d' times that of L', because Y' is equal to the preimage of X' in I. Therefore,

$$H^2 = d' L' \cdot L^2 = d'$$

where the first term is the intersection number on Y' and the second term is the intersection number on I.

STEP 6: $nd' = 2g - 2 + 2d$.

Indeed, since C has no cusps, the formula is a special case of one of the standard Plücker formulas ([18], (I,25), p. 307, see also pp. 309–310).

STEP 7: $D \equiv nH - (2g - 2 + d)F$, modulo numerical equivalence on Y'.

Indeed, D is congruent to some linear combination of H and F, because H is a section of Y'/X'. The combining coefficient of H is equal to n because of Step 3. Thus

$$D \equiv nH + aF$$

for some a. So Steps 4 and 5 yield

$$d = nd' + a.$$

Finally, Step 6 yields the assertion.

STEP 8: $K \equiv -2H + (2g' - 2 + d')F$, where K is a canonical divisor on Y'.

Indeed, since H is a section of Y'/X', there is a b such that

$$K \equiv -2H + bF.$$

In addition, also because H is a section of Y'/X', the (arithmetic) genus of $p_a(H)$ is equal to g'. Therefore,

$$2g' - 2 = (K + H) \cdot H = -d' + b.$$

STEP 9: $p_a(D) = n(g' - g) + g$.

Indeed, Steps 7 and 8 yield

$$K + D \equiv (n - 2)H + (2g' - 2g + d' - d)F.$$

Hence Steps 4 and 3 yield

$$(K + D) \cdot D = (n - 2)d + (2g' - 2g + d' - d)n = 2n(g' - g) - 2d + d'n.$$

Therefore, Step 6 yields

$$2p_a(D) - 2 = 2n(g' - g) + 2g - 2.$$

STEP 10: $p_a(D) = g = g'$.

Indeed, since $\eta\sigma : X \to D$ is birational,

$$p_a(D) \geq g.$$

And, since there is a map $X \to X'$,

$$g \geq g'.$$

Hence, the assertion follows from Step 9. (The hypothesis on g has not been used yet.)

STEP 11:   *The separable degree s of γ is equal to 1, just as Kaji's theorem asserts.*

Indeed, factor the map $X \to X'$ into

$$X \longrightarrow X^{\dagger} \longrightarrow X'$$

where the first map is separable of degree s, the second map is purely inseparable, and $X^{\dagger}$ is normal. Since the second map is purely inseparable, $X^{\dagger}$ is of genus g'. So the Hurwitz formula yields

$$2g - 2 \geq s(2g' - 2).$$

Since $g = g'$ by Step 10 and since $g \geq 2$ by hypothesis, therefore $s = 1$.

REFERENCES

[1] A. Andreotti, *On a theorem of Torelli*, Amer. J. Math. **80** (1958), 801–828

[2] V. Bayer and A. Hefez, *Strange curves*, Preprint, Spring 1989

[3] M. Deuring, *Invarianten und Normalformen elliptischer Funktionenkörper*, Math. Z. **47** (1941), 47–56

[4] A. Garcia, *The curves $y^n = f(x)$ over finite fields*, Arch. Math. **54** (1990), 36–44

[5] A. Garcia and J. Voloch, *Duality for projective curves*, Preprint, impa Série A-085-Nov./89

[6] A. Hefez, *Nonreflexive curves*, Compositio Math. **69** (1989), 3–35

[7] A. Hefez and S. Kleiman, *Notes on duality of projective varieties*, in "Geometry Today. Roma 1984," E. Arbarello, C. Procesi, E. Strickland (eds.) Prog. Math. **60**, Birkhäuser, 1985, pp. 143–184.

[8] A. Hefez and I. Vainsencher, *Strange plane curves*, Preprint, Spring 1989

[9] A. Hefez and J. Voloch, *Frobenius nonclassical curves*, Arch. Math. **54** (1990), 263–273

[10] A. Holme (with E. Lluis), *Strange varieties*, Appendix to "Embeddings, projective invariants and classifications," Monografias del Instituto de Matematicas, Univ. Nac. Autonoma de Mexico, Vol. 7, 1979

[11] M. Homma, *Funny plane curves in characteristic $p > 0$*, Comm. Alg. **15** (1987), 1469–1501

[12] M. Homma, *A souped-up version of Pardini's theorem and its application to funny curves*, Compositio Math. **71** (1989), 295–302

[13] H. Kaji, *On the tangentially degenerate curves*, J. London Math. Soc. (2) **33** (1986), 430–440

[14] H. Kaji, *On the multiple tangents of space curves*, Preprint, Waseda U., Tokyo 160, Japan, 1987

[15] H. Kaji, *On the Gauss maps of space curves in characteristic p*, Compositio Math. **70** (1989), 177–197

[16] N. Katz, *Pinceaux de Lefschetz: théorème d'existence*, SGA 7 II, Exposé XVII, Springer Lecture Notes in Math. **340** (1973), 212–253

[17] S. Kleiman, *The transversality of a general translate*, Compositio Math. **38** (1974), 287–297

[18] S. Kleiman, *The enumerative theory of singularities*, in "Real and complex singularities," P. Holm (Ed.), Proc. Conf., Oslo 1976), Sitjhoff & Noorhoof, 1977, pp. 297–396

[19] S. Kleiman, *About the conormal scheme*, in "Complete Intersections. Proc. Conf., Arcireale, Italy 1983," S. Greco, R. Strano (eds.), Lecture Notes in Math. **1092**, Springer, 1984, pp. 161-197

[20] S. Kleiman, *Tangency and duality*, in "Proc. 1984 Vancouver Conf. in Algebraic Geometry," J. Carrell, A. V. Geramita, P. Russell (eds.), CMS Conf. Proc. **6**, Amer. Math. Soc., 1986, pp. 163–226

[21] S. Kleiman (with the collaboration of A. Thorup on Section 3), *Intersection theory and enumerative geometry: a decade in review*, in "Algebraic Geometry — Bowdoin 1985," S. J. Bloch (ed.), Proc. Symposia Pure Math., Vol. **46** — Part 2, Amer. Math. Soc., 1987, pp. 321–370

[22] D. Laksov, *Indecomposability of restricted tangent bundles*, Young tableaux and Schur functors in algebra and geometry, Astérisque **87–88** (1981), 207–219

[23] E. Lluis, *Variedades algebraicas con ciertas conditiones en sus tangentes*, Bol. Soc. Mat. Mexicana (2) **7** (1962), 47–56

[24] R. Pardini, *Some remarks on plane curves over fields of finite characteristic*, Compositio Math. **60** (1986), 3–17

[25] J. Rathmann, *The uniform position principle for curves in characteristic p*, Math. Ann. **276** (1978), 565–579

[26] P. Samuel, *Courbes planes en charactéristique* 2, Séminaire P. Dubreil (1954/55), pp. 13-01–13-05

[27] P. Samuel, in "Lectures on old and new results on algebraic curves," Tata Institute of Fundamental Research, Bombay, 1966.

[28] A. Wallace, *Tangency and duality over arbitrary fields*, Proc. London Math. Soc. (3) **6** (1956), 321–342

MATHEMATICS DEPARTMENT, ROOM 2-278, M. I. T., CAMBRIDGE, MA 02139,

Contemporary Mathematics
Volume 116, 1991

Enumerative Geometry of Nonsingular Plane Cubics

STEVEN L. KLEIMAN AND ROBERT SPEISER

This paper completes the determination of the characteristic numbers for all plane cubics, begun in [CC], which treated cuspidal cubics, and continued in [NC], which treated nodal ones. The family of nonsingular plane cubics displays sufficient complexity to make its study a test case for methods aimed at higher degree curves, so we emphasize the general character of our approach. The 9-parameter family of nonsingular cubics has ten characteristic numbers, denoted by $N_{9,0}, N_{8,1}, \ldots, N_{0,9}$. By definition, each $N_{\alpha,\beta}$ counts the nonsingular cubics which pass through α general points, and are tangent to β general lines. (Because the condition to pass through a point is linear, we have $N_{9,0} = 1$. The other characteristic numbers, as we shall see, are not trivial.) These numbers were originally found independently by Maillard [M] and Zeuthen [ZC] in the 1870's, by essentially heuristic methods. Schubert discussed them in his 1879 book [Sch], so their rigorous justification, like the justification of the characteristic numbers of the singular cubics, is part of Hilbert's 15th Problem.

In our view, the characteristic numbers reflect the geometry of the closure, denoted by Γ, of the graph of the correspondence between a nonsingular cubic and its dual sextic. Indeed, in the $\mathbf{P}^9$ of all plane cubics, the divisor parametrizing the cubics tangent to a given line contains the locus of all nonreduced cubics, so that, for the first six characteristic numbers, the intersection of the corresponding divisors is proper, and it is even transversal for the first five. On Γ, however, the intersection will be proper always for general points and lines, and so we can begin. To continue, we need to normalize to obtain a smooth (but noncomplete) parameter space, so that the intersecting subschemes of codimension 1 will become divisors.

1980 *Mathematics Subject Classification* (1985 *Revision*). Primary 14N10; Secondary 14C17, 14H45.

The first author was partly supported by NSF Grant DMS-8801743.

The second author was partly supported by NSF Grant DMS-8802015.

This paper is in final form and no version of it will be submitted for publication elsewhere.

A recent paper by Aluffi [A] takes a different approach to finding the characteristic numbers of the smooth cubics: blow up $\mathbf{P}^9$ successively along smooth loci until the intersection of the 9 strict transforms becomes proper. Aluffi obtains a smooth, complete parameter space, which dominates ours birationally. His computations proceed by determining explicitly the relevant intersection calculus, recursively, at each link of the blowup chain. Earlier Sterz [St] found some of the same characteristic numbers using a general plan in some ways similar to Aluffi's. On a different tack, Miret and Xambó [MX1, MX2, MX3] have vindicated Schubert's more elaborate approach to the characteristic numbers of cuspidal cubics and nodal cubics, and they plan to do the same for smooth cubics. In addition, Aluffi [A2] and van Gastel [vanG] have, independently and somewhat differently, determined the first few nontrivial characteristic numbers of curves of higher degree.

We use a more basic parameter space than Aluffi and Sterz, but we need, as did Maillard, Zeuthen, and Schubert, the characteristic numbers of nodal cubics, which depend, in turn, on those for cuspical cubics and conics. Along the way, we learn some fascinating things about the various singular cubic curves, and, as a benefit, our calculations, at each stage, are surprisingly simple.

Our results, as do Aluffi's and Sterz's, confirm the pioneers' conclusions. Unlike theirs, our work also vindicates, and powerfully generalizes, the underlying strategy of Maillard and Zeuthen, just as the work of Miret and Xambó does that of Schubert. We do so by extending and strengthening the framework of general results about families of plane curves of any degree, with arbitrary degenerations, begun in our previous papers [CC] and [NC].

Summary. In §1, we give some basic results about nonsingular plane cubics, their dual curves, and, especially, their nodal and nonreduced degenerations. The nodal degenerations are parametrized by a locus N in the $\mathbf{P}^9$ of all cubics, an orbit under the action of the projective group, denoted $\mathbf{G}$. There are two kinds of nonreduced degenerations: unions of double lines and lines, parametrized by $U \subset \mathbf{P}^9$, and triple lines, parametrized by the Veronese surface $V \subset \mathbf{P}^9$.

Here we sound a main theme of this paper. Denote by $\mathbf{E}$ the smooth locus of the normalization of Γ. To study the characteristic numbers, we shall work on the smooth but noncomplete variety $\mathbf{E}$, a *partial compactification* of the open subset of $\mathbf{P}^9$ denoted E, which parametrizes all nonsingular plane cubics. Our plan will be to define the characteristic numbers by means of intersection products on $\mathbf{E}$, and then argue that these numbers are independent of the partial compactification $\mathbf{E}$ because they are given by 0-cycles supported only on E. Everything depends, therefore, on how much we can learn about $\mathbf{E}$ and its projection Γ.

We show in §2 that N, U and V lift to *divisors* on $\mathbf{E}$, while the loci in $\mathbf{P}^9$ parametrizing all other degenerations lift to subvarieties of higher codimen-

sion, paving the way for the intersection calculations. Further, we determine
the limit of the dual of a nonsingular cubic, as it degenerates to a curve given
by a point of N, U or V. We do this by inspecting explicit degenerations
(called *homolographies* in [**FKM**] and [**K**] because they are degenerating ho-
mologies, although often called homographies in the 19th century) and then
arguing that any degeneration looks like one of the explicit ones. (Other authors too have recently considered the degeneration of the dual of a plane
curve. Miret and Xambó devote a section of [**MX1**] to homolographic degenerations of plane curves of any degree. And van Gastel [**vanG**] and Katz
[**Kz**] consider more general degenerations, like those studied by Zeuthen in
his great monograph [**ZG**].)

A point of the pullback of N is a nodal cubic, endowed with a *double
vertex* at its node. More precisely, consider the sextic dual to a nonsingular
cubic degenerating to a nodal one: the limit sextic consists of the quartic
dual to the nodal limit curve, plus a double line dual to the node. While the
pullback of N is a copy of N, the other two pullbacks fiber nontrivially. A
point of the pullback of V, for example, is a triple line, endowed with six
unordered points (called *vertices*) which correspond to six concurrent lines in
the dual plane. Such a configuration will be called a *six-pointed triple line* in
the sequel.

In §3, we turn to intersection calculus on E. We prove two general results about 1-parameter families of plane curves, of any degree, with arbitrary
degenerations. The first, 3.1, generalizes the ramification formula [**NC**, Theorem 3.2], while the second 3.4, generalizes the coincidence formula [**NC**,
Theorem 4.2]. These give key relations in $\mathrm{Pic}(\mathbf{E})$, which we now describe.

As in our previous work, our constructions lead to three basic kinds of
divisor classes on $\mathbf{E}$. The first kind come from points and lines in $\mathbf{P}^2$. Fix
a point p, and pull back the hyperplane of plane cubics through p from $\mathbf{P}^9$
to $\mathbf{N}$. The resulting divisor class, denoted M, is independent of p. Now
choose a line l, viewed as a point in the dual plane $\check{\mathbf{P}}^2$. The sextics through
l form a hyperplane in the $\mathbf{P}^{27}$ of sextics in $\check{\mathbf{P}}^2$; its pullback defines a divisor
class M' on $\mathbf{E}$, independent of l. Classes of the second kind are given by
the divisors on $\mathbf{E}$ over the degeneracy loci N, U and V. Denote these
classes by [**N**], [**U**], and [**V**]. Divisor classes of the third kind describe how
special curve-points and their tangents move with the curve which carries
them. Here we have, for example, the class F, which represents the curves
with an inflectional tangent through a given point of $\mathbf{P}^2$.

The relations we find, 3.2 and 3.5, equate integral linear combinations of
divisors of each kind with integral linear combinations of divisors of the
other two kinds. We determine some of their coefficients in §3, using explicit
families. This process is continued in §4, where these relations are sharpened
further. We determine several key multiplicities through a close inspection,
inspired by Zeuthen's monograph [**ZG**], of the behavior of nonreduced degenerations whose limits are not too special. (Two constants remain to be found

at this stage, but they won't matter for finding the characteristic numbers.)

In §§5–7, preparing for the main results, we construct explicit models of the pullbacks of V, U and N to Γ. All three are smooth 8-folds. The first two are incidence correspondences, while the third is a copy of N. Each maps to the $\mathbf{P}^9$ of cubics and the $\mathbf{P}^{27}$ of dual sextics, and we pull back the hyperplanes corresponding to passage through a point and tangency to a line in order to compute directly the resulting characteristic numbers, which we shall need, inductively, to compute the characteristic numbers on $\mathbf{E}$. To find the characteristic numbers for the nodal degenerations, we apply our earlier results on nodal cubics [NC]. However, the map from our copy of N to $\mathbf{P}^{27}$ is different, because the dual curve is now a sextic, obtained by adding a double vertex at the node, so the characteristic numbers we obtain in this case are new.

Finally, §8, following [Z], we obtain the main results. Here, as in our previous papers, we look at the curve on $\mathbf{E}$ given by intersecting 8 divisors representing the classes M and M', show that this curve is *complete*, and then restrict to one more M or M' and reduce to the characteristic numbers of the boundary components using our relations in Pic($\mathbf{E}$).

Here there is an interesting subtlety. The boundary divisors $\mathbf{U}$ and $\mathbf{V}$ are *finite covers* of their images, denoted $\mathbf{U}_0$ and $\mathbf{V}_0$, on Γ, so the degrees of these coverings enter our calculations. Fortunately, having 10 characteristic numbers to calculate, we accumulate enough numerical information to determine these degrees: they are 1 and 40, respectively. In particular, the closed duality correspondence Γ is generically smooth along the pullback of U, but is singular, with 40 branches, along the pullback of V.

The duality correspondence. That Γ has 40 branches along the pullback of V follows easily (over $\mathbf{C}$, at least) from Hurwitz' combinatorial description [H] of the equivalence classes of simple r-sheeted covers branched at w points of the Riemann sphere. To understand the connection with the branches of Γ, consider a point of Γ over V, that is, a 6-pointed triple line. Then choose a projection center off the triple line, and join it to the 6 vertices, producing 6 concurrent lines. The nonsingular cubics tangent to all 6 lines form an algebraic family, whose irreducible components correspond both to the equivalence classes of triple covers of the sphere with 6 branch points and to the branches of Γ.

Thinking of the equations of representatives of the 40 components, it is easy to see, as Tyrrell explained in [T], that the 40 branches of Γ along the pullback of V correspond to the 40 different ways that a binary sextic form (here the discriminant of the projection from the chosen center to the triple line) can be written as the sum of a square and a cube. Tyrrell also cited a result by Clebsch [C] that there are 40 such representations, and then offered a new proof, based on enumerative techniques. In the course of his discussion, Tyrrell conjectured that the j-invariant separates the branches of Γ over a general point of V. Recently Eisenbud, Elkies, Harris, and the second author [**EEHS**] established a strong generalization of Tyrrell's conjecture for curves of any genus.

Blanket Hypotheses. As in [**CC**] and [**NC**], nearly all our results are valid over an algebraically closed based field of characteristic not 2 or 3. In the rare case that a statement requires further restriction of the characteristic, we shall say so explicitly.

1. Basics

Denote by $\mathbf{P}^9$ the space of cubic curves in $\mathbf{P}^2$. We write

$$E := \{\text{nonsingular cubics } X \subset \mathbf{P}^2\},$$

an open subscheme of $\mathbf{P}^9$. The boundary of E is a union of disjoint locally closed, reduced subschemes, each parametrizing a particular type of singular curve.

In codimension 1, we have only

$$N := \{\text{nodal cubics}\}.$$

In codimension 2, we have

$$K := \{\text{cuspidal cubics}\},$$

and

$$\Psi := \{\text{unions of a conic and a (nontangent) secant line}\}.$$

In codimension 3, we find

$$\Sigma := \{\text{unions of a conic and a tangent line}\},$$

and

$$\Delta := \{\text{unions of 3 nonconcurrent lines}\}.$$

In codimension 4, on the boundary of Δ, we find

$$\Theta := \{\text{unions of 3 concurrent lines}\}.$$

Finally, nonreduced curves appear in codimensions 5 and 7. Their parameter varieties are

$$U := \{\text{unions of a double line and another line}\},$$

and the Veronese surface

$$V := \{\text{triple lines}\}.$$

Denote by $\mathbf{P}^{27}$ the space of sextics in $\check{\mathbf{P}}^2$, and by Γ the closure of the graph of the correspondence which associates to each nonsingular cubic its dual sextic.

Denote by $\mathbf{G}$ the group $\mathrm{PGL}(2)$, acting as usual on $\mathbf{P}^2$, on the dual plane $\check{\mathbf{P}}^2$, and on $\mathbf{P}^9$ and $\mathbf{P}^{27}$ by substitution. Clearly all the subschemes of $\mathbf{P}^9$

defined above are **G**-stable. Moreover, **G** is transitive on all of them except
E. (For N and Ψ, consult [**NC**]. For K and Σ, see [**CC**]. For Δ, Θ, U,
and V, use the transitivity of **G** on four-tuples of points in general position.)
It is also clear that Γ is **G**-stable.

The partial compactification. Denote by $n(\Gamma)$ the normalization of the
duality correspondence Γ, and write **E** for the smooth locus of $n(\Gamma)$. The
G-action on Γ induces a **G**-action on **E**.

PROPOSITION 1.1. *The subschemes E and N of $\mathbf{P}^9$ lift isomorphically to*
G-*stable subschemes of* **E**.

PROOF. Denote by ϕ the canonical rational map from $\mathbf{P}^9$ to the normalization $n(\Gamma)$. Plainly, ϕ is **G**-equivariant, so the indeterminacies of this
rational map form a **G**-stable closed subscheme, denoted Z, of $\mathbf{P}^9$. Clearly
Z does not meet E. By normality, Z has codimension ≥ 2. Hence Z
cannot meet N either, so E and N lift bijectively to closed **G**-stable subschemes of the smooth set of Γ. The Proposition follows immediately.

We shall identify E with its image in **E**, which is dense and open, and
we shall regard **E** as a partial compactification of E. On the other hand,
N lifts to a divisor on **E**, denoted by **N**, a component of the boundary of
E. As we pass from $E \cup N$ to $E \cup \mathbf{N}$, each nonsingular cubic X is assigned
its dual sextic $\check{X}$ in $\check{\mathbf{P}}^2$, and, in addition, each nodal cubic X is assigned a
unique sextic $\check{X}$. For a nodal X, by [**K**, (3.11,ii)], the dual curve $\check{X}$ is the
union of the usual quartic dual to X and a double line, dual to the node of
X. Following [**Z**, **CC**, and **NC**], we say that the nodal cubic X has a *double
vertex* at its node.

The pullbacks to **E** of K, Ψ, Σ, Δ, Θ, U and V will be denoted,
respectively, by **K**, $\mathbf{\Psi}$, $\mathbf{\Sigma}$, $\mathbf{\Delta}$, $\mathbf{\Theta}$, **U** and **V**. A point t of any of these
varieties, except for **U** and **V**, parametrizes a plane curve X with isolated
singularities, together with a dual sextic, obtained by adjoining vertices as
above at appropriate singularities of X. This assignment of vertices of X
can be carried out in at most finitely many ways, so at most finitely many
points of Γ correspond to a given plane curve X. Hence at most finitely
many g in **E** correspond to a given X. Since K, Ψ, Σ, Δ, and Θ have
codimension > 1 on $\mathbf{P}^9$, we obtain the next result.

COROLLARY 1.2. *Each of the subschemes* **K**, $\mathbf{\Psi}$, $\mathbf{\Sigma}$, $\mathbf{\Delta}$ *and* $\mathbf{\Theta}$ *of* **E** *is
either empty or has codimension* > 1.

In contrast, U and V blow up to divisors on **E**. We establish this below,
after reviewing a basic construction needed for the proof.

2. Homolographies

Arbitrary plane curves. Let $X \subset \mathbf{P}^2$ denote a reduced curve of degree d;
let $c \in \mathbf{P}^2$ be a point; possibly on X, let m denote the multiplicity of X

at c, possibly 0. Let $L \in \mathbf{P}^2$ be a line off c. Denote by $\mathbf{A}^1$ the affine line $\text{Spec}(k[t])$. We first define a 1-parameter family of plane curves $\{X_t\}_{t \in \mathbf{A}^1}$ such that X is equal to X_1 and X_0 is equal to the union of the $(d - m)$-fold line and the tangent cone of X at c.

Choose homogeneous coordinates x, y, z such that $y = 0$ defines L, and such that $c = (0, 1, 0)$, at ∞ on the y-axis. Suppose that $X = V(f)$, where $f = f(x, y, z)$ is homogeneous of degree d. We define $F(x, y, z, t)$ by substituting y/t for y in $f(x, y, z)$ and then clearing denominators. Then we let

$$X_t := V(F(x, y, z, t)).$$

To study the variation of X_t, write

$$f(x, y, z) = f_d(x, z) + f_{d-1}(x, z)y + f_{d-2}(x, z)y^2 + \cdots + f_{d-e}(x, z)y^e,$$

where f_i is homogeneous of degree i in x and z. Then

$$F(x, y, z, t) = t^e f_d + t^{e-1} f_{d-1}y + t^{e-2} f_{d-2}y^2 + \cdots + f_{d-e}y^e.$$

In affine coordinates with $y = 1$, the term f_{d-e} is the leading form, so it defines the tangent cone of X_t at c for any t. First set $t = 1$: we find $X = X_1$ and so $d - e = m$. Now set $t = 0$: the affine equation $f_m = 0$ defines X_0 off L. Therefore, the rest of X_0 is $(d - m)L$. Thus, we have proved the following result.

LEMMA 2.1. *The limit curve X_0 is the union of the multiple line $(d - m)L$ and the tangent cone of X at c.*

Completing $\mathbf{A}^1$ to $\mathbf{P}^1 = \mathbf{A}^1 \cup \{\infty\}$, we can set $X_\infty = \lim_{t \to \infty} X_t$.

The defining equation of X_∞ is clearly $f(x, 0, z)$. Hence X_∞ is a union of d lines through c, joining c to the d points where X meets L. Each such line appears with the same multiplicity as the corresponding point of $X \cap L$.

A *homography* with *center c* and *target L* is defined to be the completed family $\{X_t\}_{t \in \mathbf{P}^1}$. It does not depend on the choice of coordinates. Indeed, the point $A = (x, y, z)$ corresponds to the point $B = (x, ty, z)$ and the cross ratio $(B, A, 0, \infty)$ equals t. So given A, the cross ratio, independent of the coordinates, determines B. Notice that the homography is given by the 1-parameter family of linear transformations of the plane,

$$(x, y, z) \mapsto (x, ty, z).$$

For $t \neq 0, \infty$, the corresponding linear transformation is an automorphism.

Denote by π the projection from c to L. The effect of the homography above, as $t \to 0$, is to move the points of X linearly toward L, along the fibers of π. As $t \to \infty$, the effect is to pull the points of X off L asymptotically toward the center c, as X degenerates to a union of fibers of π.

EXAMPLE 2.2. Let $X = V(y^2z - x^3 - xz^2)$. Then $c \in X$, and for $t \neq \infty$ we have

$$X_t = V(y^2z - t^2x^3 - t^2xz^2),$$

so X_0 is the doubled x-axis. In affine coordinates with $z = 1$, we see that X_∞ is the union of the 3 vertical lines $x = -i$, 0, i.

EXAMPLE 2.3. Let $X = V(x^2z - y^3 - yz^2)$. Then $c \notin X$, and for $t \neq \infty$ we have

$$X_t = V(t^3x^2 - y^3 - t^2yz^2),$$

so X_0 is the tripled x-axis. In affine coordinates with $z = 1$, we see that $X\infty$ is the union of the line at ∞ with the doubled y-axis.

In the dual plane, homolographies are reversed. More precisely, denote by $\check{c}$ the line in $\check{\mathbf{P}}^2$ dual to $c \in \mathbf{P}^2$, and by $\check{L} \in \check{\mathbf{P}}^2$ the point dual to L. Choose coordinates in $\check{\mathbf{P}}^2$ which are dual to x, y, z. The following result is easily checked by examining the action on the coordinates in the dual plane.

LEMMA 2.4. *As X moves in the homolography $\{X_t\}$ with center c and target L, the dual curve $\check{X}$ moves in the homolography $\{\check{X}_u\}$ with center $\check{L}$ and target $\check{c}$, where the parameters satisfy $tu = 1$.*

Nonreduced degenerate cubics. Fix a target L, and a nonsingular plane cubic X. Now we allow the center c to be anywhere in $\mathbf{P}^2 - L$, and we investigate the dependence on c of the resulting homolography $\{X_t\}$. We are equally interested in the limiting behavior of the dual curve $\check{X}_t$. Thus we take the limit, not in $\mathbf{P}^9$ or $\mathbf{P}^{27}$, but in Γ. Denote by e_0 the corresponding limit point as $t \to 0$.

PROPOSITION 2.5. *For a nonsingular plane cubic X, moving in a homolography as above, we have:*

(1) If c is off X, then e_0 parametrizes the tripled line L, equipped with 6 vertices, which can be any distinct points on L (that is, any 6 distinct points arise from some X).

(2) If $c \in X$, but c is not a flex, then e_0 represents the union of the doubled line L and the tangent line, denoted T, to X at c. There is a double vertex at $T \cap L$, and 4 additional vertices elsewhere on L. These four vertices can be any distinct points, other than the double vertex, on L.

(3) If c is a flex of X, then e_0 again represents the union of $2L$ and T, but now there is a triple vertex at $T \cap L$, and 3 other vertices, which can be arbitrary but distinct points of L.

PROOF. (1) By Lemma 2.1, we know that the limit cubic is a triple line. So, using 2.4, look at the dual curve $\check{X}_u$ as $u \to \infty$. We obtain a union of 6 lines, because $\deg(\check{X}_u) = \text{class}(X) = 6$. Because homolographies are given by linear transformations of the plane that fix c and the points of L, each X_t, in particular X, is tangent to the 6 lines. So now take 6 general lines in $\mathbf{P}^2$, concurrent at c. Tangency to each imposes at most one condition on a smooth cubic. Counting dimensions, we see that the set of smooth cubics tangent to all 6 lines, but not passing through c, is certainly nonempty. Specializing the lines, by semicontinuity, we can still find an X as needed. Taking a homology with this X, we can realize an arbitrary set of 6 distinct points on L as vertices. Hence (1) holds.

With $c \in X$, Lemma 2.1 gives $X_0 = 2L \cup T$ as claimed. Denote by p the point where L and T meet. For the vertices, we look in $\check{\mathbf{P}}^2$ again. When c is not a flex, the general $\check{X}_u$ is simply tangent to the line $\check{p}$ at the point $\check{T}$. Hence, as $u \to \infty$, we obtain the double line from $\check{T}$ to the center $\check{L}$. This double line corresponds to a double vertex at p. Again, counting dimensions shows that every configuration of vertices appears, so (2) is true. For (3), with c a flex, the line $\check{P}$ is now a cuspidal tangent of $\check{X}_u$. The 3 intersections at the cusp give a triple vertex as claimed. The rest of the argument goes as before.

COROLLARY 2.6. *The pullbacks of U and V to Γ are irreducible 8-folds. A general point of either pullback can be represented as the limit of a suitable homology.*

PROOF. Choose v in the pullback of V, and denote by $L \subset \mathbf{P}^2$ the corresponding triple line. The sextic in $\check{\mathbf{P}}^2$ corresponding to v must obviously be a union of 6 lines meeting at the point of $\check{\mathbf{P}}^2$ dual to L, and these 6 lines in the dual plane give 6 vertices on L. By 2.5(1), every configuration consisting of a line equipped with six vertices appears, so the pullback of V to Γ is an 8-fold, whose general point represents the limit of a homology. Since the space of all configurations in $\mathbf{P}^2$ consisting of a line equipped with 6 unordered points is irreducible, so is the pullback of V. Similar reasoning treats the pullback of U.

Finally, we pass to $\mathbf{E}$.

PROPOSITION 2.7. *The pullbacks $\mathbf{U}$ and $\mathbf{V}$ are nonempty divisors on $\mathbf{E}$.*

PROOF. The variety $n(\Gamma)$ is smooth on the $\mathbf{G}$-stable open set $\mathbf{E}$, which contains all points of codimension 1. Hence, by 2.6, dense open subsets of U and V lift to nonempty divisors on $\mathbf{E}$.

3. Ramification and coincidence formulas

First we recall the general ramification formula [NC, Theorem 3.2] for a family of plane curves.

Suppose given an r-parameter family $\{X_t \mid t \in T\}$ of curves in $\mathbf{P}^2$, where the parameter scheme $\mathbf{T}$ is irreducible and smooth, and such that the general X_t is reduced, irreducible, and reflexive. We have a natural dual family, denoted by $\check{X}_t$, in which the dual $\check{X}_t$ of the general X_t moves. Denote by d the degree of the X_t and by d' the degree of the $\check{X}_t$, that is, the class of X_t.

We define divisor classes M and M' on T as follows. As before, write $\mathbf{P}^N$ for the space of curves of degree d in $\mathbf{P}^2$. Dually, write $\mathbf{P}^{N'}$ for the space of curves of degree d' in $\check{\mathbf{P}}^2$. We have natural maps $f: T \to \mathbf{P}^N$ and $g: T \to \mathbf{P}^{N'}$, defined by the family X_t. Let H (resp. H') denote the divisor class of a hyperplane in $\mathbf{P}^N$ (resp. $\mathbf{P}^{N'}$). We set

$$M = f^*(H) \quad \text{and} \quad M' = g^*(H'),$$

in $\mathrm{Pic}(T)$. Hence M represents the divisor T which parametrizes the X_t through a general point of $\mathbf{P}^2$, while M' represents the divisor parametrizing the X_t tangent to a general line in $\mathbf{P}^2$.

The ramification formula [NC, Theorem 3.2, p. 170] is this identity of divisor classes on T:

(3.1) $$(2d - 2)M = M' + \sum_i s_i[S_i] + \sum_j u_j[U_j].$$

Each S_i is reduced and irreducible, and it parametrizes curves X_t with a given type of isolated singularity falling on the given line. The coefficients s_i and u_j are positive integers. For ordinary nodes, $s_i = 2$, and for ordinary cusps, $s_i = 3$. The contributions of multiple components, for example double and triple lines, are more subtle, for they depend on the family. Each U_j is also reduced and irreducible, and it parametrizes curves X_t with nonreduced components of a given type. The formula is functorial if the multiplicities of the components of the pullbacks of the S_i and U_j are duly taken into account.

For good families of nonsingular cubics, double and triple lines will typically appear. In the next result, we apply Formula 3.1 to obtain two relations of divisor classes on $\mathbf{E}$, in a preliminary form. These will yield two of Zeuthen's key numerical formulas [Z, (1) and (2), p. 727].

Denote by $\{X_t\}$ the pullback to $\mathbf{E}$ of the canonical family of cubics on $\mathbf{P}^9$. We define $\mathbf{F}$ to be the class of the divisor E, as in Formula 3.1, on $\mathbf{E}$ parametrizing the curves $\check{X}_t$ in the dual family, such that a cusp of $\check{X}_t$ lies on a general line in $\check{\mathbf{P}}^2$. By duality, $\mathbf{F}$ is the class of the divisor on $\mathbf{E}$ parametrizing the curves $X_t \subset \mathbf{P}^2$ for which an inflectional tangent passes through a general point of $\mathbf{P}^2$. More precisely, we work with the cycle of cusps (resp. flexes) as in [NC, §1, p. 159].

COROLLARY 3.2. *For the canonical family of cubics on* $\mathbf{E}$, *there are positive integers* A, B *and* C *such that we have in* $\mathrm{Pic}(\mathbf{E})$

$$4M = M' + A[\mathbf{U}] + B[\mathbf{V}] \quad \text{and} \quad 10M' = M + [\mathbf{N}] + C[\mathbf{U}] + 3[\mathbf{F}].$$

PROOF. For the first identity, let $\{X_t\}$ be the canonical family of cubics on $\mathbf{E}$. The first formula follows directly from Formula 3.1; the class $[N]$ does not appear, because the node of a general nodal cubic does not lie on any given line. For the second identity, take the dual family $\{\check{X}_t\}$. In 3.1, components S_i parametrizing ordinary cusps on the general line count 3 times, it follows that the coefficient of [F] in the second formula is 3. To complete the proof, we only need to verify that [N], which corresponds to the locus parametrizing the curves $\check{X}_t$ containing double lines, appears with multiplicity 1. Because $\mathbf{N}$ has exactly one branch over its image N in $\mathbf{P}^9$, we can check that the multiplicity in question is equal to 1 by restricting directly to a specific family, and we do so next in Example 3.3.

EXAMPLE 3.3. Choose any nodal plane cubic N. Because the nodal cubics are an orbit for $\mathbf{G}$, we can choose coordinates in $\mathbf{P}^2$ such that N is the folium of Descartes, that is,

$$y^2 = x^3 - x^2.$$

To vary N in the simplest possible way, set

$$F = y^2 - x^3 + x^2 + t.$$

The locus of $F = 0$ on $\mathbf{P}^2 \times \mathbf{A}^1$, where $A^1 = \mathrm{Spec}(k[t])$, defines a family of curves $\{X_t\}$, whose general member is a nonsingular cubic, and whose special member X_0 is N. To work out the dual family, we substitute $mx + b$ for y and take the discriminant in x.

We find that the dual family is defined, in $\mathbf{A}^2 \times \mathbf{A}^1 = \mathrm{Spec}(k[m, b, t])$, by a sextic polynomial $G(m, b, t)$, such that $G(m, b, 0)$ is divisible by b^2 but not by b^3. Explicitly, we have

$$\begin{aligned}
G(m, b, t) = {} & m^6 t + 3m^4 t + 3m^2 t + t \\
& + m^4 b^2 + 2m^2 b^2 + 9mbt + b^2 + (27/4)t^2 + 9m^3 bt \\
& + 9mb^3 + (27/2)b^2 t + m^3 b^3 + (27/4)b^4,
\end{aligned}$$

where the denominators make sense because the characteristic is not 2.

Hence, as expected, when $t \to 0$, the dual curve $\check{X}_t$ approaches a limit sextic $\check{X}_t$, which is the union of the double line defined by $b^2 = 0$ and the quartic dual to N in the usual sense. Further, denote by $G_b(m, b, t)$ the derivative with respect to b. To determine the contribution of the double line in $\check{X}_0$ to the ramification divisor, we can proceed as in [NC, §3, case (3)]. We first restrict to a line Λ: $m = const.$, and inspect the ramification locus, denoted R, of the projection $(m, b) \mapsto (m, 0)$, as a subscheme of $\Lambda \times \mathbf{A}^1$. Hence, assign the weight 0 to m, and the weight 1 to b and t. By the cited analysis, R is defined by the vanishing of both $G(m, b, t)$ and its derivative $G_b(m, b, t)$. From the explicit computation of G, we find that the leading form of G is

$$(m^6 + 3m^4 + 3m^2 + 1)t,$$

and that of G_b is

$$(2m^4 + 4m^2 + 2)b + (9m^3 + 9m)t.$$

The only intersection point of $V(G)$ and $V(G_b)$ in $\Lambda \times \mathbf{A}^1$ relevant to the contribution of the double line is $(b, t) = (0, 0)$. Here, the curves $V(G)$ and $V(G_b)$ are smooth and intersect transversally, provided that $2m^4 + 4m^2 + 2$ is nonzero (the characteristic is not 2!). It follows immediately that the double line in $\check{X}_0$ counts once in the ramification formula. This completes the verification of Theorem 3.2.

Our next result generalizes the coincidence formula [NC, 3]. Its restriction to a good one-parameter family, whose general member is a reduced, irreducible plane curve of any degree, appears as item (9) in Zeuthen's list of basic enumerative relations for plane curves of arbitrary degree [ZG, p. 44]. Its restriction to a good one-parameter family whose general member is a nonsingular cubic is formula (3) in [Z], as well as item (9) in Zeuthen's list [ZG, p. 82] of basic formulas for nonsingular cubics.

To state our result, denote by $\{X_t \mid t \in T\}$ and r-parameter family of curves in $\mathbf{P}^2$, satisfying the hypotheses of formula (3.1). As before, we write d for the degree and d' for the class of the general X_t.

Let $\mathbf{A} \subset \mathbf{P}^2 \times \mathbf{P}^2 \times \check{\mathbf{P}}^2 \times T$ denote the closure of the set of (x, y, L, t) such that x is a smooth point of X_t, and L is the tangent line to X_t at x, and $y \in X_t \cap L$ but $x \neq y$. Let Φ be the locus $x = y$ on $\mathbf{A}$, and let ε be the divisor on Φ defined by the condition that L go through a general point of $\mathbf{P}^2$. Denote the irreducible components of codimension 1 of the image of ε in T by A_1, A_2, and so on.

THEOREM 3.4. *Under the assumptions above, we have an identity in* $\mathrm{Pic}(T)$ *of the form,*

$$(d - 2)M' + (d + d' - 4)M = \sum_i a_i[A_i],$$

for suitable positive integers a_i.

PROOF. The argument is essentially the same as for [NC, Theorem 4.2], except now the degree and class are general. Let $\mathbf{X}$ be the total space of the family, and let

$$\alpha \colon \mathbf{A} \to C\mathbf{X},$$

be the projection to the conormal scheme. Note that the degree of α is $d - 2$, because the general tangent to the general X_t is ordinary, as X_t is reflexive by hypothesis.

As on page 177 of [NC], we have the coincidence formula,

$$[\Phi] = x + y - L,$$

where, by abuse of notation, x, y, and L denote the divisor classes on $\mathbf{A}$ corresponding to the conditions that x (resp. y) lies on a given line and that L goes through a given point. Multiplying by L, we obtain

$$[\varepsilon] = xL + yL - L^2.$$

Using Schubert's incidence formula as on page 177, we obtain

$$[\varepsilon] = x^2 + y^2 + L^2.$$

(In [NC], the first displayed formula for ε was marred by a small typographical error.)

To prove the asserted formula, we push down the last formula to T. It suffices to show (1) that x^2 pushes down to $(d-2)M$, (2) that y^2 pushes down to $(d'-2)M$, and (3) that L^2 pushes down to $(d-2)M'$.

Since $\alpha: \mathbf{A} \to C\mathbf{X}$ is of degree $d-2$ and since $C\mathbf{X} \to \mathbf{X}$ is birational, pushing down x^2 to $C\mathbf{X}$, we obtain $(d-2)\xi^2$, where ξ is the locus where the point x of $\mathbf{P}^2$ is fixed. By [NC, 5.6, p. 248], the class ξ^2 pushes down to M. Thus we obtain (1). Dually, we obtain (3).

Finally, to prove (2), define a new map $\mathbf{A} \to \mathbf{X}$ by assigning to (x, y, L, t) the pair (y, t). This map is of degree $d'-2$ because for a general pair (y, t) there are exactly $d'-2$ tangents L to X_t through y as X_t is reflexive. Now, applying the projection formula for this map $\mathbf{A} \to \mathbf{X}$, we find by [NC, 5.6, p. 248] that (2) holds.

COROLLARY 3.5. *For the canonical family of cubics on* $\mathbf{E}$, *there are integers* d *and* e *such that we have*

$$M' + 5M = [\mathbf{F}] + d[\mathbf{U}] + e[\mathbf{V}],$$

in Pic($\mathbf{E}$), *where* $\mathbf{F}$ *is the reduced divisor of* $t \in T$ *such that a flex tangent of* X_t *passes through a given point* c.

PROOF. Use c to define ε. Then the image of ε on $\mathbf{E}$ has three components: $\mathbf{U}$, $\mathbf{V}$, and $\mathbf{F}$, where $\mathbf{F}$ is the closure of the set of $t \in T$ such that X_t has a flex tangent (at a smooth point) through P. It remains to prove that $\mathbf{F}$ appears with multiplicity 1 on the right side of the formula in 3.5. We do this, as usual, by reducing to a special family.

EXAMPLE 3.6. (To test the multiplicity of $\mathbf{F}$.) Denote by C a given nonsingular plane cubic, and choose affine coordinates x and y in $\mathbf{P}^2$, such that C has a flex at the origin (denoted O) with the x-axis as flex tangent. To verify 3.5, we also need to choose C and its flex to represent a smooth point of the divisor of pairs such that the flex tangent goes through a given point. Because C is tangent to the x-axis at the origin, the coordinate x restricts to a local parameter, denoted t, at O on C.

Now let a point P move on C. Using a variant of the method of [NC, Example 4.4, p. 177], we define a family of plane curves $\{C_P\}$, parametrized by $P \in C$ near O, such that $C_O = C$, such that a general C_P has the x-axis as an ordinary tangent at the origin. To do so, we use the standard inner product in the (x, y)-plane, as in [NC, loc. cit.], to set up an "orthonormal frame" at each $P \in C$, with coordinates X and Y, such that the X-axis is tangent to C at P. Writing C in the frame coordinates X and Y, we obtain C_P.

In this family, as $P \to O$, a flex of C_p approaches the origin as its flex tangent approaches the X-axis. We denote by Q the third point at which the cubic C_p meets the X-axis. As $P \to O$, we know that $Q \to P$, and that this coincidence gives a point of ε over F. Our goal is to show that this coincidence counts exactly once, so that F will appear with multiplicity one on the right side of the identity of 3.5.

Denote by L the tangent to C at P. For the family $\{C_p\}$, consider the associated incidence variety $\mathbf{A}$. To show that F counts once in 3.5, we need to show that the curve traced by (P, Q, L, t) on $\mathbf{A}$ meets the exceptional locus E once. Because P and L are constant in the frame coordinates, it is easy to see that we can reach our goal if we can show that the X-coordinate of Q vanishes to first order in the local parameter t.

To verify the last assertion, we consider the analytic family given by C_p, over the formal neighborhood of 0 on the parameter curve C. Identify the completed local ring $\widehat{\mathscr{O}}_{C,P}$ with $k[[t]]$. The analytic family is then given by the equation

$$(*) \qquad F(X, Y, t) = f(X, Y) + tG(X, Y, t) = 0,$$

where $f(X, Y) \in k[x, y]$ defines C, and $G \in k[[t]][X, Y]$. Now C has a flex at the origin, with the X-axis as flex tangent. Hence, rescaling X if necessary, we have

$$(**) \qquad\qquad f(X, 0) = X^3.$$

In particular the monomials X and X^2 do not appear in $f(X, Y)$. Further, we have chosen our frame so that the general C_p has an ordinary tangent with the X-axis at the origin. Hence we must have

$$(***) \qquad\qquad G(X, 0, t) = -X^2 \varphi(X, 0, t),$$

for some $\varphi \in k[[t]][X, Y]$. Therefore we find

$$F(X, 0, t) = X^2(X - t\varphi(X, 0, t)).$$

Denote by $\alpha(t) \in k[[t]]$ the X-coordinate of Q, represented by a series in t. To complete the verification, it suffices to show that $\alpha(t)$ vanishes to first order in t, because the point P and the line L are fixed in the frame coordinates X, Y. Since $\alpha(t)$ is a nonzero root of $F(X, 0, t)$, we know that $X - \alpha(t)$ divides $X - t\varphi(X, 0, t)$ in the factorial ring $k[[t]][X]$. Therefore, if we can show that $\varphi(X, 0, t)$ is not divisible by t, it will follow that $\alpha(t)$ vanishes to first order in t.

By our construction, the frame coordinate X differs from x by a series of the form $t + \cdots$. Using $(*)$ and $(**)$, we find

$$
\begin{aligned}
G(X, 0, 0) &= \left.\frac{\partial}{\partial t}\right|_{t=0} F(X, 0, t) \\
&= \left.\frac{\partial}{\partial t}\right|_{t=0} f(X + t + \cdots, t^3 + \cdots) \\
&= \left.\frac{\partial}{\partial t}\right|_{t=0} (X + t + \cdots)^3 - t^3 + \cdots \\
&= 3X^2.
\end{aligned}
$$

Because the characteristic is prime to 3, it follows that $G(X, 0, t)$ is not divisible by t. Therefore, by $(***)$, the factor $\varphi(X, 0, t)$ is also not divisible by t, as we had to show.

REMARKS 3.7. (1) The proof above works in all characteristics but 3, and applies equally well to cuspidal and nodal cubics C. In particular, we recover [NC, 4.5, p. 179].

(2) A second verification of 3.6, which also works in any characteristic but 3, can be obtained by using the group structure on the elliptic curve C. The group structure can be defined, in any characteristic, by the classical method of taking line sections. (In fact, any elliptic curve in $\mathbf{P}^2$, whatever the characteristic, can be brought into the Weierstrass form described in [Ta].) Denote the X-coordinate of q by $\xi(t)$. To verify that $\xi(t)$ vanishes to first order in t, we reason as follows:

By our construction, $x - X$ is represented by a series of the form $\alpha(t) = t + \cdots$, so it suffices to check that the x-coordinate of q is represented by a series $\beta(t) = at + \cdots$, where $a \neq 1$, so that t doesn't cancel out of $\xi(t) = \beta(t) - \alpha(t)$. In fact, we shall show in any characteristic that $a = -2$, which differs from 1 in all characteristics but 3.

To determine a, consider the group structure on the elliptic curve C with identity point 0, a flex. The assignment $p \mapsto q$ is given by multiplication by -2; we denote by $f : C \to C$ the resulting isogeny. The derivative of f is multiplication by -2, because f induces multiplication by -2 in Lie (C). (Using the functorial isomorphism between Lie (C) and the set of base point preserving maps $\operatorname{Spec}(k[\varepsilon]/(\varepsilon^2)) \to C$, this is immediate.)

To complete the verification, compose f with the projection to the x-axis, to obtain $g : X \to \mathbf{A}^1$. We have $g(0) = 0$, and $g(p)$ is the x-coordinate of q. By the chain rule, the derivative of g is multiplication by -2, so $\beta(t) = -2t + \cdots$, as was to be shown.

(3) The method of (2) can also be used for cuspidal and nodal C, which are group varieties, by the same construction, away from their singular points. In particular, this gives still another proof for [NC, 4.5, p. 179].

4. General degenerations

For many families which appear in practice, the results in Corollaries 3.2 and 3.5 can be improved. In this section we shall be concerned with an arbitrary family $\{X_t\}$ of plane curves, as in the last section, but subject to the following conditions:

(1) the general X_t is a nonsingular cubic;

(2) the family $\{X_t\}$ is the pullback of the canonical family on $\mathbf{E}$ via a map $f: T \to \mathbf{E}$;

(3) for t in a dense subset of $f^{-1}(U)$, the curve X_t has four distinct vertices on its double line; and

(4) for t in a dense of $f^{-1}(V)$, the curve X_t has six distinct vertices on its triple line.

A family satisfying (1)–(4) will be said to have *general degenerations*. It is conceivable that the canonical family on $\mathbf{E}$ may not be an example. However, there is a maximal $\mathbf{G}$-stable open subscheme $\mathbf{E}_{\mathrm{gen}}$ containing $\mathbf{N}$ and dense open subsets of $\mathbf{U}$ and $\mathbf{V}$ that is an example.

Consider, for another example, curves T which are the intersections of 8 divisors on $\mathbf{E}$, such that each divisor intersects each $\mathbf{G}$-orbit properly. Then, by proper intersection arguments based on [$\mathbf{S}$, §1], even though there are infinitely many orbits, general translates of our divisors will intersect in a curve $T \subset \mathbf{E}$ which satisfies our 4 conditions.

Elementary systems. For a point $p \in \mathbf{P}^2$, the divisor $\mathscr{D}_p$ on E parametrizing the curves through p will satisfy our conditions. We define $\mathscr{D}_p$ to the direct image on $\mathbf{E}$ of the pullback of p to the total space of the canonical family. For a line $l \in \check{\mathbf{P}}^2$ we define the divisor $\mathscr{E}_l$ dually. Hence $\mathscr{E}_l$ parametrizes the curves tangent to l.

Let α and β be nonnegative integers such that $\alpha + \beta = 8$. An *elementary system of nonsingular cubics*, denoted $C_{\alpha, \beta}$ corresponding to α and β is defined to be a 1-parameter family of curves parametrized by the intersection of α general translates of $\mathscr{D}_p$ and β general translates of $\mathscr{E}_l$. Hence $C_{\alpha, \beta}$ consists the curves (from $\mathbf{E}$) through α general points, tangent to β general lines. By the reasoning above or by that in [**FKM**], each elementary system has general degenerations.

One-Parameter Families. We now investigate the contributions of multiple line components to the first ramification formula of 3.2 and to the coincidence formula 3.5, for a family $\{X_t\}$ of plane cubics with general degenerations. As explained in [**NC**, §3], we can restrict attention to 1-parameter families. Our analysis is based on that of Zeuthen's monograph [**ZG**, Tredie Afsnit, §§41–51], which also treats curves of higher degree.

First we establish some notation. Let $\{X_t\}_{t \in T}$ be the given 1-parameter family, whose general member is a nonsingular plane cubic, and whose parameter curve T is smooth. We fix a base point $0 \in T$, and denote by t a local parameter in a neighborhood of $0 \in T$. We shall assume that the curve

X_0 contains a multiple line, and we shall choose coordinates in $\mathbf{P}^2$ such that the multiple line is defined by the vanishing of y^2 or y^3. We shall work in affine coordinates x, y, and write $F(x, y, t) = 0$ for a local defining equation of the total space of the family.

First we consider the case of a double line in X_0. The defining polynomial F in suitable coordinates is either of the form

$$A_1 y^2 + B_3 t + \varphi t^2,$$

or of the form

$$A_1 y^2 + 2B_2 yt + C_3 t^2 + \varphi t^3.$$

Here each coefficient A_i, B_j, and C_3 is a polynomial in x and y, not divisible by y, with a subscript indicating its maximum possible degree. The term denoted by φ is a polynomial in of x, y and t, not divisible by t. Following Zeuthen, we shall use small letters a, b to denote the values of the corresponding coefficients when $y = 0$.

It is easy to see that vertices other than the root of a_1 on the double line are given by roots of the y-discriminant of F, as $t \to 0$, and that these vertices have the same multiplicities which they do as roots of the discriminant. In the first case, we find at least a triple vertex (rather than the double vertex which appears in general) at the intersection of the two components $V(A_1)$ and $V(y^2)$, because the limit discriminant $4a_1 b_3$ has at most 3 roots away from the root of a_1. Indeed, we can choose coordinates so that no vertex lies at infinity on the double line. Hence this family does not have general degenerations.

In the second case, the discriminant limits to a unit times $b_2^2 - a_1 c_3$, which in general, defines 4 distinct vertices on the double line. By 2.5 (2), every configuration of distinct vertices appears in this way. Clearly, in our situation, we can collect terms so that neither B_2 nor C_3 involve y.

Now, when we restrict to a vertical line $\Lambda: x = \mathrm{const.}$, we find that $F|\Lambda$ has a quadratic leading form,

$$A_1 y^2 + 2B_2 yt + C_3 t^2,$$

while the leading form of the derivative F_y is

$$2A_1 y + 2B_t t.$$

Hence, on $\Lambda \times T$, for a general Λ, the intersection of $V(F)$ and $V(F_y)$ at the origin is double. Therefore, by the analysis of [NC, §3], the double line in X_0 contributes *two* ramifications in the first formula of 3.2. (In the first case, a special degeneration, the same method, by the way, gives one ramification.)

It follows that double lines count twice in the first formula of 3.2, if the family has general degenerations, because a multiplicity of one happens only for special degenerations.

Next we consider triple lines in the second formula of 3.2. Here we can make the usual change of variables to eliminate y^2, and write

$$F = y^3 + B_2 yt^\alpha + \varphi t^{\alpha+\beta},$$

where again φ is a polynomial. The y-discriminant of this cubic is

$$\mathfrak{d} = 4B^3 t^{3\alpha} - 27\varphi^2 t^{2(\alpha+\beta)}.$$

Here there are two cases to consider.

(1) When $\alpha \neq 2\beta$, the limit of the discriminant locus, as $t \to 0$, is defined either by the vanishing of b^3 or φ^2. In either case we obtain multiple vertices, so the family does not have general degenerations. Hence, if there are distinct vertices, then we must have $\alpha = 2\beta$.

(2) When $\alpha = 2\beta$, we obtain 6 distinct vertices if the discriminant is general. Indeed, use (1) noting by 2.5(1) that homologations (where $\beta = 1$) give every possible configuration of distinct vertices. Notice that it follows from this analysis that the general sextic in one variable is the sum of a square and a cube in at least one way. (We shall see later that the number of essentially distinct such representations is 40, partly on the basis of the multiplicity we are about to compute. Compare [T].)

Continuing in the case $\alpha = 2\beta$, we see that by replacing t by an appropriate root, we have reduced to considering the case when $\alpha = 2$ and $\beta = 1$, as for a homology. Now, following the method of the previous case, we restrict to a general line $\Lambda: x = \text{const.}$, and inspect the leading forms of F and its partial derivative F_y. These are respectively cubic and quadratic in x and t, both products of linear factors. But there can be no common linear factor. Indeed, the vertical line Λ is general, so if there were a common factor, the original polynomial $F(x, y, t)$ would be reducible, contradicting the irreducibility of the general X_t. We conclude that the multiplicity of the triple line is 6, in the first formula of 3.2, for a family with general degenerations.

Finally we turn to the coincidence formula 3.5, continuing under the notations above. If we choose coordinates so that the base point c, as in the discussion before Theorem 3.4, is at ∞ on the y-axis, then it's clear that vertical tangents (that is, through c) contribute coincidences in the formula 3.5 as they approach distinct vertices on a triple line. Hence there are 6 such coincidences per triple line, in a family with general degenerations. We claim that each such coincidence counts once.

Indeed, we can test with a homology from a general c, in fact from any c from which the given smooth cubic X has 6 distinct tangents, for then we have a general degeneration as $t \to 0$. By definition of the family, each point p of tangency, and the third point q where the line pc meets X, move linearly toward their collision on the triple line. It follows immediately that each of the 6 coincidences on the triple line counts once.

Based on these 1-parameter families, our usual method of restriction yields the following result about the maximal family with general degenerations because $\text{Pic}(\mathbf{E}_{\text{gen}})$ and $\text{Pic}(\mathbf{E})$ are equal by virtue of Corollary 1.2.

THEOREM 4.1. *For the pullback to* $\mathbf{E}$ *of the canonical family of plane cubics, we have the relations,*

(1) $$4M = M' + 2[\mathbf{U}] + 6[\mathbf{V}],$$

(2) $$10M' = M + [\mathbf{N}] + C[\mathbf{U}] + 3[\mathbf{F}],$$

and

(3) $$M' + 5M = [\mathbf{F}] + D[\mathbf{U}] + 6[\mathbf{V}],$$

in $\mathrm{Pic}(\mathbf{E})$, *for suitable positive integers* C *and* D, *where* $\mathbf{F}$ *is the reduced divisor of* $t \in T$ *such that a flex tangent of* X_t *passes through a given point* c.

5. Characteristic numbers: degenerate curves over V

We begin with preliminaries about the incidence correspondences which will serve as building blocks for our constructions, both in this section and the next.

The incidence correspondences. Denote by $\mathbf{I}$ the incidence correspondence of points and lines in $\mathbf{P}^2$. Write α for the pullback of $c_1(\mathcal{O}_{\mathbf{P}^2}(1))$ to $\mathbf{I}$. We have $\mathbf{I} = P(\mathcal{E})$, where $\mathcal{E} := \Omega^1_{\mathbf{P}^2}(-1)$, and the embedding of $\mathbf{I}$ in $\mathbf{P}^2 \times \check{\mathbf{P}}^2$ is given by the map from $\mathcal{E}$ in the standard exact sequence,

$$0 \to \mathcal{E} \to \mathcal{O}_{\mathbf{P}^2}^{\oplus 3} \to \mathcal{O}_{\mathbf{P}^2}(1) \to 0.$$

Hence the tautological class $\mathbf{I} = P(\mathcal{E})$ is the pullback, denoted β, of $c_1(\mathcal{O}_{\check{\mathbf{P}}^2}(1))$ to $\mathbf{I}$. We also have [**F**, p. 189, and Example 8.34, p. 141] the standard relation

(5.1) $$\alpha\beta = \alpha^2 + \beta^2.$$

The Chern polynomial $c_t\mathcal{E} = 1/(1 + \alpha t) = 1 - \alpha t + \alpha^2 t^2$ gives $c_1\mathcal{E} = -\alpha$ and $c_2\mathcal{E} = \alpha^2$. Using [**F**, Example 3.2.6, p. 57], we easily obtain the Chern classes of the symmetric powers $S^k\mathcal{E}$. In particular, we have

(5.2) $$c_1 S^4\mathcal{E} = -10\alpha, \qquad c_2 S^4\mathcal{E} = 55\alpha^2,$$

and

(5.3) $$c_1 S^6\mathcal{E} = -21\alpha, \qquad c_2 S^6\mathcal{E} = 231\alpha^2.$$

Denote by $\mathbf{I}^{(k)}$ the kth symmetric power $P(S^k\mathcal{E})$ of $\mathbf{I}$ over $\mathbf{P}^2$. A point of $\mathbf{I}^{(k)}$ can be viewed as a configuration in $\mathbf{P}^2$ consisting of k lines through a given intersection point. This time we have projections from $\mathbf{I}^{(k)}$ to $\mathbf{P}^2$ and to the $\mathbf{P}^N$ of curves of degree k in $\mathbf{P}^2$. Again write α for the pullback of the hyperplane class on $\mathbf{P}^2$, and now write β for the pullback of the hyperplane class on $\mathbf{P}^N$. (In particular, α stands for the condition that the intersection point lie on a given line, while β represents the condition that the union of k concurrent lines goes through a given point.)

Using (5.1)–(5.3), we obtain the rational equivalence rings

(5.4)
$$A\mathbf{I}^{(4)} = \mathbf{Z}[\alpha, \beta]/(\alpha^3, \beta^5 - 10\alpha\beta^4 + 55\alpha^2\beta^3),$$
$$A\mathbf{I}^{(6)} = \mathbf{Z}[\alpha, \beta]/(\alpha^3, \beta^7 - 21\alpha\beta^6 + 231\alpha^2\beta^5).$$

In the first ring, the class of a point is $\alpha^2\beta^4$, while in the second it is $\alpha^2\beta^6$.

We now dualize the preceding discussion. Viewing $\check{\mathbf{I}}$ as a scheme over the *dual* plane $\check{\mathbf{P}}^2$, we now write $\mathbf{I}^{(k)}$ for the k-fold symmetric product over $\check{\mathbf{P}}^2$. It parametrizes the *k-pointed lines* in $\mathbf{P}^2$. Denote by $\mathbf{P}^{N'}$ the space of curves of degree k in $\check{\mathbf{P}}^2$; we have a natural map $\check{\mathbf{I}}^{(k)} \to \mathbf{P}^{N'}$, assigning to each k-pointed line the reducible curve in $\check{\mathbf{P}}^2$ whose components are the concurrent lines dual to the k given points in $\mathbf{P}^2$.

For an integer $d \geq 0$, denote by $\mathbf{P}^N$ the space of plane curves of degree d. To parametrize the *k-pointed d-fold lines* in $\mathbf{P}^2$, we map $\check{\mathbf{I}}^{(k)}$ to $\mathbf{P}^N$ by the d-fold Veronese map. Pulling back the universal family on $\mathbf{P}^N$, we obtain a universal family of k-pointed d-fold lines. Pulling back the universal family on $\mathbf{P}^{N'}$, we obtain a universal family of dual k-pointed d-fold lines as well. Equipped in this way, $\check{\mathbf{I}}^{(k)}$ will be denoted $\mathbf{I}_{d,k}$.

For a point p (resp. a line l) in $\mathbf{P}^2$, denote by $\widetilde{D}$ (resp. $\widetilde{E}$) the class, in $\mathrm{Pic}(\mathbf{I}_{d,k})$, of the pullback of the hyperplane $H_p \subset \mathbf{P}^N$ (resp. $H_l \subset \mathbf{P}^{N'}$) parametrizing the k-pointed d-fold lines through p (resp. tangent to l). In particular, a k-pointed d-fold line passes through p if its d-fold line does, and is tangent to l whenever l passes through one of its k points.

The incidence variety $\mathbf{V}_0$. Set

$$\mathbf{V}_0 := \mathbf{I}_{3,6},$$

an 8-fold. The projection $\mathbf{V}_0 \to \mathbf{P}^9$ corresponds to the restriction, to $V = \{\text{triple lines}\}$, of the projection $\Gamma \to \mathbf{P}^9$ of the duality correspondence defined in §1. The projection $\mathbf{V}_0 \to \mathbf{P}^{27}$ assigns to each point of $\mathbf{V}_0$ the reducible sextic corresponding to the 6 vertices. It follows that the image of V on the duality correspondence Γ identifies with a dense open subset of $\mathbf{V}_0$.

The *characteristic numbers* here are $V_{8,0}, \ldots, V_{0,8}$:

$$V_{\alpha, \beta} := \int_{\mathbf{V}_0} \widetilde{D}^\alpha \widetilde{E}^\beta, \qquad (\alpha + \beta = 8).$$

Taking general translates, we find set-theoretically that only when $\alpha \leq 2$ can $V_{\alpha, \beta}$ be nonzero. In the relations in the second line of (5.4), $\widetilde{D}$ corresponds to 3β, and $\widetilde{E}$ corresponds to α. Computing in the rational equivalence ring, we easily obtain

PROPOSITION 5.5. *The canonical family of 6-pointed triple lines on* $\mathbf{V}_0$ *has characteristic numbers*

$$V_{8,0} = \cdots = V_{3,5} = 0, \quad V_{2,6} = 9, \quad V_{1,7} = 63 \quad and \quad V_{0,8} = 210.$$

PROOF. We determine $V_{1,7}$ to illustrate the method. We have

$$V_{1,7} = \int \tilde{D}\tilde{E}^7 = \int 3\beta\alpha^7$$
$$= 3\int \beta(21\alpha^6\beta - 231\alpha^5\beta^2) = 63\int \alpha^6\beta^2$$
$$= 63,$$

using the *duals* of the relations in the second line of (5.4).

These results agree with Maillard's and Zeuthen's.

6. Characteristic numbers: degenerate curves over U

Continuing the notation of the last section, write W for $\mathbf{I}_{2,4}$ a 6-fold. Equipped with suitable extra structure, W will help us to describe explicitly the part of Γ over $U \subset \mathbf{P}^9$, the space of unions of lines and double lines.

For the extra structure, denote by B the blowup of $\check{\mathbf{P}}^2 \times \check{\mathbf{P}}^2$ along its diagonal. Hence [KO, IV B, pp. 368–369] the variety B parametrizes ordered triples (L, M, x), where L and M are lines, and $x \in L \cap M$ is a point. Equivalently, B is the fiber product of I with itself over $\mathbf{P}^2$. Denote also by L (resp. x) the map $B \to \check{\mathbf{P}}^2$ (resp. $B \to \mathbf{P}^2$) induced by the projection $(L, M, x) \mapsto L$ (resp. $(L, M, x) \mapsto x$). We also have a map $B \to \mathbf{P}^9$, obtained as follows. Write v for the Veronese embedding $\check{\mathbf{P}}^2 \to \mathbf{P}^5$, where the target is the space of plane conics. Use $v \times \mathrm{id}$ first to map B to $\mathbf{P}^5 \times \check{\mathbf{P}}^2$, then the Segre embedding followed by a projection to map into the $\mathbf{P}^9$ of plane cubics. This composite map sends the triple (L, M, x) to the reducible cubic whose cycle is $2L + M$. We have

$$A(B) = \mathbf{Z}[\beta_1, \beta_2, \alpha]/(\alpha^3, \beta_i^3, \alpha^2 + \beta_i^2 - \alpha\beta_i), \qquad (i = 1, 2),$$

where β_i denotes the pullback of $c_1(\mathscr{O}_{\check{\mathbf{P}}^2}(1))$ from the ith factor $\check{\mathbf{P}}^2$, and α denotes the pullback $x^*c_1(\mathscr{O}_{\mathbf{P}^2}(1))$. The class of a point of B is:

$$[\mathrm{pt}_B] = \alpha^2\beta_1\beta_2 = \alpha\beta_1^2\beta_2 = \alpha\beta_1\beta_2^2,$$

as one checks easily using the description of B as a fiber product, together with the relations in $A(B)$. We also have

$$\alpha^2\beta_i^2 = 0,$$

by the relations.

The incidence variety $\mathbf{U}_0$. To construct our model for the part of Γ over U, we view B (resp. W) as a scheme over $\check{\mathbf{P}}^2$, via the map L (resp. the structure map). Finally, we define $\mathbf{U}_0$ to be the fiber product:

$$\mathbf{U}_0 = B \times_{\check{\mathbf{P}}^2} W.$$

Tracing through the constructions, it's easy to see that U_0 parametrizes configurations in $\mathbf{P}^2$ consisting of a single line, a double line, and 4 unordered points on the double line.

We map U_0 to the $\mathbf{P}^9$ of plane cubics via the projection $B \to \mathbf{P}^9$ above. We map U_0 to the $\mathbf{P}^{27}$ of dual sextics by viewing the 4 points as vertices, and throwing in a double vertex at x by means of the twofold Veronese $\mathbf{P}^2 \to \mathbf{P}^5$, as before. This time the composite is

$$U_0 \to \mathbf{P}^2 \times (\check{\mathbf{P}})^4 \to \mathbf{P}^5 \times (\mathbf{P}^2)^4 \to \mathbf{P}^{27},$$

and the pullbacks of $c_1(\mathcal{O}_{\mathbf{P}^2}(1))$ and $c_1(\mathcal{O}_{\check{\mathbf{P}}^2}(1))$ to U_0 are easy to identify.

Equipped with its projections to $\mathbf{P}^9$ and $\mathbf{P}^{27}$, it is clear that U_0 parametrizes all configurations consisting of the union of a line and a double line, with 4 simple vertices on the double line and one double vertex at a given point on the intersection of the line and the double line. In particular, the image of U on Γ identifies with an open dense G-stable subset of U_0.

Putting the last results together, we find that the intersection ring of U_0 is

(6.1) $$A(U_0) = \mathbf{Z}[\alpha, \beta, \mu, \psi]/R,$$

where R denotes the ideal generated by the relations:

$$
\begin{aligned}
\alpha^5 &= 10\alpha^4\beta - 55\alpha^3\beta^2, \\
\beta\psi &= \beta^2 + \psi^2, \\
\mu\psi &= \mu^2 + \psi^2, \\
\beta^3 &= \mu^3 = \psi^3 = 0.
\end{aligned}
$$

(with (R) labelling the block)

Here α represents the condition that one of the 4 vertices on the double line L lies on a given line, 2β represents the condition that the double line L goes through a given point, μ represents the condition that the single line M goes through a given point, and 2ψ represents the condition that the double vertex x lies on a given line.

Using the relations (R), we see directly that the class of a point of U_0 is

$$[\mathrm{pt}_{U_0}] = \alpha^4[\mathrm{pt}_B] = \alpha^4\beta^2\mu^2 = \alpha^4\beta^2\mu\psi = \alpha^4\beta\mu^2\psi = \alpha^4\beta\mu\psi^2,$$

and that we have

$$\beta^2\psi^2 = \mu^2\psi^2 = 0.$$

Further, denote by $\tilde{D}$ (resp. $\tilde{E}$) the class in $\mathrm{Pic}(U_0)$ of the pullback of $\mathcal{O}_{\mathbf{P}^9}(1)$ (resp. of $\mathcal{O}_{\mathbf{P}^{27}}(1)$). Tracing through the maps as before, we obtain

$$\tilde{D} = 2\beta + \mu, \qquad \tilde{E} = \alpha + 2\psi.$$

The *characteristic numbers* here are $U_{8,0}, \ldots, U_{0,8}$:

$$U_{\alpha,\beta} := \int_{U_0} \tilde{D}^\alpha \tilde{E}^\beta, \qquad (\alpha + \beta = 8).$$

From the explicit description (6.1), by direct calculation, we obtain the following result.

PROPOSITION 6.2. *The Characteristic numbers of the canonical family of degenerate nonsingular cubics on* $\mathbf{U}_0$ *are:*

$$U_{8,0}, \dots, U_{0,8} = 0, 0, 0, 0, 24, 240, 885, 1470, 0.$$

PROOF. To illustrate the method, we compute $U_{3,5}$. We have

$$U_{3,5} = \int \widetilde{D}^3 \widetilde{E}^5 = \int (2\beta + \mu)^3 (\alpha + 2\psi)^5$$

$$= \int (12\beta^2 \mu + 6\beta\mu^2)(\alpha^5 + 10\alpha^4\psi + 40\alpha^3\psi^2),$$

using the relations $\psi^3 = \beta^3 = \mu^3 = 0$. Then, using the relation on α^5, we obtain

$$\int (12\beta^2\mu + 6\beta\mu^2)(10\alpha^4\beta - 55\alpha^3\beta^2 + 10\alpha^4\psi + 40\alpha^3\psi^2)$$

$$= \int [120\alpha^4\mu\beta^2\psi + 480(\beta^2\mu\psi^2) + 60\alpha^4\beta^2\mu^2 + 60\alpha^4\beta\mu^2\psi$$

$$+ 240\alpha^3(\beta\mu^2\psi^2)].$$

In the last expression, the terms in parentheses are pullbacks of classes on the 4-fold B, so they vanish. Each intersection product in the remaining terms gives the class of a point, so we find $U_{3,5} = 240$, as claimed.

These results agree with Maillard's and Zeuthen's.

7. Characteristic numbers: degenerate curves over N

By the results of [NC, Theorem 5.13], the characteristic numbers of nodal plane cubics are

(7.1) $\quad N_{8,0}, \dots, N_{0,8} = 12, 36, 100, 240, 480, 712, 756, 600, 400.$

As before, we denote by $\mathbf{N}$ the locally closed subscheme of $\mathbf{E}$ parametrizing the nodal degenerations.

For a point $p \in \mathbf{P}^2$, we denote by H_p the hyperplane in $\mathbf{P}^9$ parametrizing the cubics through p. It is easy to see that $\mathscr{D}_p$, defined in §4, is the pullback to $\mathbf{E}$ of the divisor H_p under the natural map $\mathbf{E} \to \mathbf{P}^9$. Under the identification of $\mathbf{N}$ with the open set of nodal cubics in the partial compactification of [NC], our present $\mathscr{D}_p \cap \mathbf{N}$ corresponds to the divisor denoted by D_p in [NC]. It parametrizes the nodal degenerations through p. Denote by D the class of $\mathscr{D}_p \cap \mathbf{N}$ in $\mathrm{Pic}(\mathbf{N})$, and by d the class of D_p. (Letting the projective group $\mathbf{G}$ act on $\mathbf{P}^2$, it follows immediately that D and d do not depend on p.) Clearly we have

(7.2) $\qquad\qquad\qquad\qquad D = d.$

Now choose a line l in $\mathbf{P}^2$. Denote by H_l the hyperplane in $\mathbf{P}^{27}$ parametrizing the sextics in $\check{\mathbf{P}}^2$ through l. Then it is easy to see that $\mathscr{E}_l$, defined in §4, is the pullback of H_l under the natural map $\mathbf{E} \to \mathbf{P}^{27}$. Restricting, we obtain a map $\mathbf{N} \to \mathbf{P}^{27}$ that factors as follows. Compose the node map

$$\mathbf{N} \to \mathbf{P}^2 := \{\text{lines in } \check{\mathbf{P}}^2\},$$

with the double Veronese embedding to obtain a map

$$\mathbf{N} \to \mathbf{P}^5 = \{\text{conics in } \check{\mathbf{P}}^2\}.$$

This, together with the map $\mathbf{N} \to \mathbf{P}^{14}$, which assigns the quartic dual to the underlying nodal cubic, gives a map $\mathbf{N} \to \mathbf{P}^5 \times \mathbf{P}^{14}$. Follow with the Segre embedding into $\mathbf{P}^{89}$, and project into $\mathbf{P}^{27}$. Hence the pullback of H_l to $\mathbf{N}$ is represented (modulo linear equivalence) by the sum of twice the pullback of (i) the class of the nodal cubics with node on l, and (ii) the class of the nodal cubics tangent to l. Denote the pullbacks of these classes by b and e, respectively. Write E for the class of $\mathscr{E}_l \cap \mathbf{N}$ in $\mathrm{Pic}(\mathbf{N})$. Then we obtain

$$(7.3) \qquad\qquad E = 2b + e.$$

Denote by $\widehat{N}_{8,0}, \ldots, \widehat{N}_{0,8}$ the *characteristic numbers* of the canonical family on $\mathbf{N}$. To be precise, for nonnegative integers α, β such that $\alpha + \beta = 8$, we set

$$\widehat{N}_{\alpha,\beta} = \int_{\mathbf{N}^+} D^\alpha E^\beta.$$

Here $\mathbf{N}^+$ denotes any compactification of $\mathbf{N}$ which is a $\mathbf{G}$-space: by the standard proper intersection argument, the characteristic number is represented by cycles in $\mathbf{N}$, so the choice of compactification doesn't matter.

PROPOSITION 7.4. *The canonical family of nodal degenerations on* $\mathbf{N}$ *has characteristic numbers*

$$\widehat{N}_{8,0}, \ldots, \widehat{N}_{0,8} = 12, 42, 192, 1168, 2784, 8832, 21828, 38072, 50448.$$

PROOF. The case of $\widehat{N}_{6,2}$ will illustrate the method. Making the substitutions (7.2) and (7.3), we have

$$\widehat{N}_{6,2} = \int d^6(2b+e)^2 = \int (d^6 e^2 + 4d^6 eb + 4d^6 b^2)$$
$$= N_{6,2} + 4b(C_{6,1} + 4N_{6,0}^{np}.$$

Here, following the notation of [NC], the elementary system of nodal cubics through 6 points, tangent to one line is denoted $C_{6,1}$ and $b(C_{6,1})$ denotes the number of curves in $C_{6,1}$ with nodes on a given line. We find $b(C_{6,1})$ from the (known) characteristic numbers of $C_{6,1}$ by the formula $4\mu = \mu' + 2b$, which follows from [NC, Theorem 3.1(1)]. Here μ (resp. μ') is the characteristic number $N_{7,1}$ (resp. $N_{6,2}$) of nodal cubics given in (7.1). Finally, $N_{6,0}^{np}$ denotes the number of nodal cubics through 6 points with a node at a given point. These characteristic numbers are given in [NC, Corollary 6.6], and we find

$$\widehat{N}_{6,2} = 100 + 4(22) + 4(1) = 192.$$

The other computations are similar.

8. Characteristic numbers: smooth cubics

Numerical Invariants. Suppose given a one-parameter family $\{X_t\}_{t \in T}$, where T is a smooth, complete curve, given by a map $T \xrightarrow{\phi} \mathbf{E}$. For a divisor class D on $\mathbf{E}$, because the curve T is complete, we can take the degree $\int_T \phi^* D$, a nonnegative integer. We apply this to the divisor classes M, M', $[\mathbf{N}]$, $[\mathbf{U}]$, $[\mathbf{V}]$ and $[\mathbf{F}]$, obtaining numerical invariants μ, μ', n u, v and f for the given family. The first two invariants are the *characteristic numbers* of the family: μ (resp. μ') counts the number of $t \in T$ such that X_t j passes through a given point (resp. is tangent to a given line) in $\mathbf{P}^2$. The third invariant, n, counts the number $t \in T$ such that X_t has its node on a given general line, the invariant u (resp. v) counts the number of t such that X_t is given by a point of U (resp. of V), and the last invariant, f, counts the number of t such that a flex tangent goes through a given general point.

Pulling back the relations of 4.1 to T and integrating, we obtain numerical relations:

(8.1)
$$4\mu = \mu' + 2u + 6v,$$
$$10\mu' = \mu + n + Cu + 3f,$$
$$\mu' = 5\mu = f + Du + 6v.$$

Characteristic numbers, Elementary systems. Fix a point p and a line l in $\mathbf{P}^2$, and recall that the divisors $\mathscr{D}_p$ and $\mathscr{E}_l$ on $\mathbf{E}$ defined in §4, parametrize the nonsingular cubics (and suitable degenerations) which pass through p and are tangent to l, respectively. Their classes, denoted M and M', do not depend on p and l, because of the transitive action of $\mathbf{G}$ on the points and lines of $\mathbf{P}^2$. For nonnegative integers α and β, such that $\alpha + \beta = 9$, we define the *characteristic number* $N_{\alpha, \beta}$ to be

$$\int_{\mathbf{E}^+} M^\alpha (M')^\beta \cap [\mathbf{E}^+],$$

where the integral is taken over any compactification $\mathbf{E}^+$ of $\mathbf{E}$ on which the projective group $\mathbf{G}$ acts, for example $n(\Gamma)$.

We claim that this integral represents an intersection on the locus $E \subset \mathbf{E}$ of nonsingular cubics, and therefore does not depend on the compactification. Indeed, it is obvious that any $\mathscr{D}_p$ (resp. $\mathscr{E}_l$) meets each of the infinitely many $\mathbf{G}$-orbits on $\mathbf{E}$ property. Hence, by [S, §1], for α general $\mathscr{D}_p$ and β general $\mathscr{E}_l$, the intersection cycle cannot lie in any 8-fold boundary component of E in $\mathbf{E}$. (Alternatively, we could argue as in [FKM].) So the integral is well-defined, and in addition, each $N_{\alpha, \beta}$ counts only nonsingular cubics.

Now suppose that $\alpha + \beta = 8$. The intersection scheme of α general $\mathscr{D}_p$ and β general $\mathscr{E}_l$ will form a *complete* curve in $\mathbf{E}$, because, by [S, loc. cit.], we can avoid any subset of the compactification having codimension > 1, in particular any boundary component of $\mathbf{E}$, in its compactification. (Indeed, by 1.2, we can avoid the boundary as well as the singular locus of $\mathbf{E}$ in the compactification $n(\Gamma)$, so we can do so in any compactification.) Denote this complete curve in $\mathbf{E}$ by $T_{\alpha,\beta}$, and equip it with the pullbacks from $\mathbf{E}$ of the canonical families of cubics and dual sextics.

Although the parameter curve $T_{\alpha,\beta}$ may not be reduced, we normalize the components of its underlying cycle and define numerical invariants as above, by linearity, as in [CC] and [NC], so that the identities (8.1) hold. The resulting structure, roughly speaking, is an integral linear combination of 1-parameter families satisfying our general hypotheses. We shall call it the *elementary system* $S_{\alpha,\beta}$ of nonsingular plane cubics. Now, we may calculate!

The characteristic numbers of an elementary system. For an elementary system $S_{\alpha,\beta}$, here with $\alpha + \beta = 8$, we have $\mu = N_{\alpha+1,\beta}$ $\mu' = N_{\alpha,\beta+1}$, and $n = \widehat{N}_{\alpha,\beta}$, the last because $\mathbf{N}$ has degree 1 over $N \subset \mathbf{P}^9$ by 1.1, hence over the pullback of N to Γ. Denote by A (resp. B) the degree of $\mathbf{U}$ (resp. $\mathbf{V}$) over $\mathbf{U}_0$ (resp. over $\mathbf{V}_0$). For any $S_{\alpha,\beta}$, let's write u_0 (resp. v_0) for the characteristic number $U_{\alpha,\beta}$ (resp. $V_{\alpha,\beta}$), as found in the last section. Because the divisor classes used to define the characteristic numbers in every case were pullbacks of the hyperplane classes $\mathbf{P}^9$ and $\mathbf{P}^{27}$, they are obviously compatible with restriction, so we have

$$u = Au_0, \quad \text{and} \quad v = Bv_0.$$

In particular, our first relation becomes

$$(8.2) \qquad\qquad 4\mu = \mu' + 2Au_0 + 6Bv_0.$$

This relation is equivalent to Zeuthen's formula [Z, (3), p. 727]. Substituting for u, and v in the remaining relations, eliminating f, and then eliminating μ' from (8.2), we obtain a relation equivalent to Zeuthen's formula [Z, (3), loc. cit]:

$$(8.3) \qquad\qquad 12\mu = n + Fu + Gv,$$

where

$$F = A(C - 3D + 14), \qquad G = 24B.$$

Now, taking the elementary systems one at a time, beginning with $S_{8,0}$, we shall compute the characteristic numbers $E_{\alpha,\beta}$ for nonsingular plane cubics.

When $\alpha > 4$, nonreduced degenerations don't appear ($u_0 = v_0 = 0$) as we saw in §7. Because the condition to pass through a point a linear, we have $N_{9,0} = 1$. Hence the system $S_{8,0}$ has $\mu = 1$, so (8.2) gives $\mu' = N_{8,1} = 4$. Continuing, we find

$$N_{\alpha,9-\alpha} = 4^{9-\alpha} \quad \text{for } \alpha = 9, 8, 7, 6, 5.$$

Now look at $S_{4,4}$. We have $\mu = N_{5,4} = 256$, by the last case, and $n = 2784$, $u_0 = 24$ and $v_0 = 0$ by 7.4, by 6.2, and by 5.5. Using (8.3), we find

$$F = 12,$$

and (8.2) gives

$$\mu' = N_{4,5} = 1024 - 48.A.$$

Next, consider $S_{3,5}$. Here $\mu = 1024 - 48A$, while $n = 8832$, $u_0 = 240$ and $v_0 = 0$. This time, using $F = 12$, in (8.3), we obtain

$$A = 1.$$

The expression for μ and (8.2) give

$$\mu = N_{4,5} = 976, \qquad \mu' = N_{3,6} = 3424.$$

For $S_{2,6}$, we have $\mu = 3424$, while $n = 21828$, $u_0 = 885$ and $v_0 = 9$. Here (8.3) gives $G = 960$, so it follows that

$$B = 40.$$

Using this in (8.2), we find

$$\mu' = N_{2,7} = 9766.$$

For $S_{1,7}$, we have $\mu = 9766$, while $n = 39072$, $u_0 = 1470$ and $v_0 = 63$. Here (8.2) alone gives

$$\mu' = N_{1,8} = 21004.$$

Finally, for $S_{0,8}$, we find

$$\mu' = N_{0,9} = 33616,$$

again by (8.2).

THEOREM 8.4 (Zeuthen [Z, pp. 727–729]; Maillard [M, p. 51]). *The characteristic numbers $N_{\alpha,\beta}$ for nonsingular cubics are*:

$$N_{9,0}, \cdot, N_{0,9} = 1, 4, 16, 64, 256, 976, 3424, 9766, 21004, 33616.$$

If the characteristic is 0, for any α, β, each solution curve counts exactly once in $N_{\alpha,\beta}$.

PROOF. Only the last assertion needs to be checked, so let the characteristic be 0. We argue as in [FKM]. (It would be nice to have an alternative proof based on [S].) Denote by $\prod$ the product of α copies of $\mathbf{P}^2$ and β copies of $\check{\mathbf{P}}^2$, and let π denote a point of $\prod$. Define $\mathfrak{X}$ to be closure in $\mathbf{E} \times \prod$ of the space of all $(t, p_1, \ldots, p_\alpha, l_1, \ldots, l_\beta)$, for $t \in \mathbf{E}$, with the p_i on the corresponding curve X_t, and the l_j tangent to X_t at smooth points. We let Y denote the pullback of π of $\mathfrak{X}$. Because $\mathbf{G}^9$ acts transitively on $\prod$, properness of the general translate ([KT] or [S]) shows that Y has dimension 0. Because the characteristic is 0, transversality of the general translate (same references) shows that Y is reduced. It now follows from the construction that the projection $Y \to \mathbf{E}$ is injective, with image the intersection of divisors given by the coordinates of π, and this proves the theorem.

COROLLARY 8.5. *The degree of* $\mathbf{U}$ *over* $\mathbf{U}_0$ *is* 1. *The degree of* $\mathbf{V}$ *over* $\mathbf{V}_0$ *is* 40.

PROOF. This follows from the computation preceding the statement of 8.4.

REMARK 8.6. In characteristic 0, it follows from 8.4 that the divisors $\mathscr{D}_p$ and $\mathscr{E}_l$ are generically reduced. In fact, this holds in any characteristic. Indeed, for $\mathscr{D}_p$, the pullback of a hyperplane in $\mathbf{P}^9$, this is trivial. For $\mathscr{E}_l$, we can restrict to the boundary of E in $\mathbf{E}$, and check that the pullback of H_l is generically reduced on each component, and we can do this by inspecting the corresponding map to $\mathbf{P}^{27}$ directly. In each case this map is obtained by projecting the Segre embedding of a suitable product, so we reduce to checking on each factor. For example, on $\mathbf{N}$, one factor is the dual map to $\mathbf{P}^{14}$ on the space of nodal cubics, and the pullback of a hyperplane here is generically reduced by [NC, 5.12]. The other factor is a Veronese, corresponding to the double vertex at the node, so the pullback is generically reduced for general l because the general conic is reduced. For U and V, we reduce immediately to the incidence varieties U_0 and V_0, in the second case by avoiding the branch locus of the projection. On each incidence variety, it's easy to inspect directly.

REFERENCES

[A] P. Aluffi, *The characteristic numbers of smooth plane cubics*, (A. Holme and R. Speiser, eds.) Proc. 1986 Sundance Conference, Lecture Notes in Math., vol. 1311, 1988, pp. 1–8.

[A2] ____, *Two characteristic numbers for smooth plane curves of any degree*, Preprint.

[C] A. Clebsch, *Zür Theorie der binären Formen sechster Ordnung und der Dreitheilung der hyperelliptischen Funktionen*, Abh. Braunschweig. Wiss. Ges. **14**, 1–59.

[EEHS] D. Eisenbud, N. Eikies, J. Harris, and R. Speiser, *On the Hurwitz scheme and its monodromy*, Proc. 1989 Zeuthen Symposium (S. Kleiman and A. Thorup, eds.) (to appear).

[F] W. Fulton, *Intersection theory*, New York and Heidelberg, 1984.

[FKM] W. Fulton, S. L. Kleiman, and R. MacPherson, *About the enumeration of contacts*, Proc. Conf. on open problems in algebraic geometry, Ravello 1982 (C. Ciliberto, F. Ghione, and F. Orecchia, eds.), Lecture Notes in Math., vol. 997, 1983, pp. 156–196.

[vanG] L. van Gastel, *An excess intersection theoretic approach to the characteristic numbers of plane curves*, to appear in the Proc. 1989 Zeuthen Symposium (S. Kleiman and A. Thorup, eds.) (to appear).

[H] A. Hurwitz, *Über Riemann'sche Fläche mit gegebenen Verzweigungspunkten*, Math. Ann. **39** (1891), 1–61.

[Kz] S. Katz, *Formal discriminants and limits of conormal schemes of families of plane curves*, 1989 Zeuthen Symposium, (S. Kleinman and A. Thorup, eds.) (to appear).

[KO] S. L. Kleiman, *The enumerative theory of singularities*, Real and complex singularities, (P. Holm, Sijthoff, and Noordhoff, eds.) (1977), 297–396.

[KT] ____, *Transversality of the general translate*, Compositio Math. **28** (1973), 287–297.

K ____, *About the conormal scheme*, Proc. 1983 Arcireale Conf., Lecture Notes in Math., vol. 1092, 1984, Springer-Verlag, Berlin and New York, pp. 161–197.

[CC] S. L. Kleiman and R. Speiser, *Enumerative geometry of cuspidal plane cubics*, Vancouver Proc., Canad. Math. Soc. Proc., vol. 6, Providence, 1986, 227–268.

[NC] ____, *Enumerative geometry of Nodal plane cubics*, Proc. 1986 Sundance Conf., Lecture Notes in Math., vol. 1311, Springer-Verlag, Berlin and New York, 1988, pp. 156–196.

[M] S. Maillard, *Récherche des charactéristiqués des systèmes élémentaires de courbes planes de troisième ordre*, Cusset, Paris, 1871.

[MX1] J. Miret and S. Xambó Descamps, *On Schubert's degenerations of cuspidal cubics*, 1987 Sitges conference, Enumerative Geometry, (S. Xambó Descamps, ed.), Lecture Notes in Math. (to appear).

[MX2] ____, *On the geometry of cuspidal plane cubics*, Proc. 1988 Trento conference, Algebraic Curves and Projective Geometry, (E. Ballico and C. Ciliberto, eds.) (to appear).

[MX3] ____, *On the geometry of nodal plane cubics* I: *the condition p* , Proc. 1989 Zeuthen Symposium, (S. Kleiman and A. Thorup, eds.) (to appear).

[Mu] D. Mumford, *Abelian varieties*, Tata Inst. Studies, Oxford and Bombay, 1970.

[Sch] H. Schubert, *Kalkül der abzählenden Geometrie*, Teubner, Leipzig (1879), reprint, Springer-Verlag, Berlin and New York, 1979.

[S] R. Speiser, *Transversality theorems for families of maps*, Proc. 1986 Sundance Conf., (A. Holme and R. Speiser, eds.), Lecture Notes in Math., vol. 1311, Springer-Verlag, Berlin and New York, 1988, pp. 235–254.

[S2] ____, *Limits of conormal schemes*, these proceedings.

[St] U. Sterz, *Berührungsvollständigung für ebene Kurven dritter ordnung*, I, II, III, IV, Beiträge Algebra Geom. **16** (1983), 45–68; **17** (1984), 115–150; **20** (1985), 161–184; **21** (1986), 91–108.

[T] J. A. Tyrrell, *Degenerate plane cubics and a theorem of Clebsch*, Bull. London Math. Soc. **5** (1973), 203–208.

[Ta] J. Tate, *The arithmetic of elliptic curves*, Invent. Math. **23** (1974), 179–206.

[Z] H. Zeuthen, *Détermination des charactéristiques des systèmes élémentaires de cubiques*, C. R. Acad. Sci. Paris **74** (1872), 726–729.

[ZG] ____, *Almindelige Egenskaber ved Systemer af plane Kurver*, Acad. Roy., Copenhagen, Nat. and Math. Sci. **10** (1873), 287–393.

MATHEMATICS DEPARTMENT, ROOM 2-278, MASSACHUSETTS INSTITUTE OF TECHNOLOGY, CAMBRIDGE, MA 02139

DEPARTMENT OF MATHEMATICS, BRIGHAM YOUNG UNIVERSITY, PROVO, UT 84602

Contemporary Mathematics
Volume **116**, 1991

The Gaussian Map for Certain Planar Graph Curves

RICK MIRANDA

1. Introduction and the statement of the main theorem

Let G be a cubic graph, i.e., a graph whose vertices each have valence three. Let v be the number of vertices and e the number of edges of G. Then the assumption that G is cubic gives the relation $3v = 2e$.

One can define a unique stable curve X_G from the cubic graph G as follows. X_G is the union of rational curves, one for each vertex of G, and they are joined according to the edges of G to form ordinary nodes. It is easy to see that X_G is a stable curve of genus $g = (v/2) + 1$; it is called the *graph curve* associated to G. We will denote by C_v the component of X corresponding to the vertex v of G, and by P_e the node of X corresponding to the edge e of G.

Given any smooth curve X, one has a natural map

$$(1.1) \qquad \phi : \wedge^2 H^0(X, \omega_X) \to H^0(X, \Omega^1_X \otimes \omega^2_X),$$

locally defined by sending $f(z)dz \wedge g(z)dz$ to $[f(z)g'(z) - g(z)f'(z)](dz)^3$, in terms of a local parameter z on X (see [**W**]); it is well-defined, as the reader can easily check i.e., it is independent of the choice of local parameter. Moreover, it can be defined (using essentially the same formula) for stable curves (see [**C-H-M**]): if ω_X is invertible, and has ω as a local generator, then $\phi(f\omega \wedge g\omega) = [fdg - gdf] \cdot \omega^2$.

We will refer to this map as the Gaussian map for X; it is, for a canonical curve X, the map on global sections for the Gauss map for X. In [**C-H-M**], it was referred to as the Wahl map for X.

For a general curve of genus g equal to 10 or at least 12, the Gaussian map ϕ is surjective (this is the content of [**C-H-M**]). This is an open condition in the moduli space for curves of genus g, and so it suffices to find one curve of a given genus with ϕ surjective to prove this. In [**W**], J. Wahl shows that

---

1980 *Mathematics Subject Classification* (1985 *Revision*). Primary 14H10, 14H45.
This paper is in final form and no version of it will be submitted for publication elsewhere.

most complete intersections have ϕ surjective, but not all genera are covered this way; in [C-H-M], the examples found covering all genera equal to 10 or at least 12 are graph curves.

At the other extreme, it is easy to see that the dimension of the cokernel of ϕ is $3g - 2$ for a hyperelliptic curve of genus g. Moreover, this is the maximum corank of ϕ (see [C-M1]). It is an interesting question to determine the possible coranks of ϕ for stable curves of genus g, and to investigate the stratification of the moduli space $\mathcal{M}_g$ by this corank.

It is the purpose of this note to prove that for certain graph curves obtained from planar cubic graphs (with some extra hypotheses given precisely below), the Gaussian map ϕ has corank one. This is interesting because such a graph curve occurs naturally in projective space as the hyperplane section of a union of 2-planes, which is a degenerate form of a K3 surface; if one could prove that this union of 2-planes could be smoothed to a K3 surface, then one could verify, using the results of [W], that for the general curve lying on a K3 surface, the Gaussian map ϕ would have corank one. Hence the general point of the "corank of $\phi = 1$" locus in $\mathcal{M}_g$ would be exactly the curves lying on K3's; this is the kind of description of the stratification of $\mathcal{M}_g$ mentioned above that one desires. The reader should consult [C-M1] and [C-M2] for information concerning these applications.

To explain the conditions under which the result holds, we require some definitions.

Let G be a planar cubic graph, i.e., G is a graph all of whose vertices have valence three, and such that an embedding of G into the 2-sphere has been fixed. G then decomposes the sphere into v vertices, e edges, and f faces; we take these to be the closed sets in question. Note that we still have $3v = 2e$, and by Euler's theorem we have $v - e + f = 2$. This implies that $v = 2g - 2$, $e = 3g - 3$, and $f = g + 1$, for an integer g; g is the genus of the graph curve associated to G.

A face is an n-gon if there are n edges around the perimeter of the face. Note that a 1-gon is the interior of a loop, and a 2-gon is the region between two edges joining the same two vertices. We have some special notation in case n is small: a 3-gon is a *triangle*, and a 4-gon is a *square*.

The *edge-neighborhood* of a face is the face plus all its vertices plus all of the edges adjacent to all of its vertices. This includes of course the edges adjacent to the face, plus the edges "sticking out" from the face.

Given an edge, the faces *adjacent to* the edge are the (one or) two faces containing that edge on their perimeter (Figure 1):

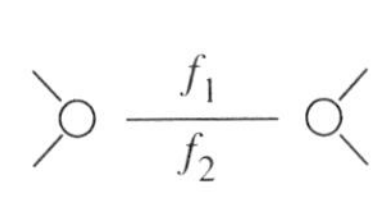

FIGURE 1

FIGURE 2

The faces *neighboring* the edge are the other two (Figure 2).
The main theorem can now be stated.

1.2. THEOREM. *Let G be a planar cubic graph, and assume the following.*

(a) *For every edge e of G, the faces adjacent to e are different and have only e and the vertices of e in common.*

(b) *For every edge e of G, the faces neighboring e are different, and have only e in their common edge-neighborhoods.*

Then the Gaussian map for the graph curve X_G associated to G has corank 1.

1.3. REMARKS. (a) Note that these assumptions imply that the decomposition of the 2-sphere determined by G has no n-gons with $n \leq 4$. In particular, G has no loops or multiple edges.

(b) If f_n is the number of n-gons in the decomposition of the sphere, then Euler's theorem can be restated as $\sum f_n(6 - n) = 12$. Therefore, since $f_n = 0$ for $n \leq 4$ by remark (a), we must have $f_5 \geq 12$, hence $f \geq 12$, so $g \geq 11$.

(c) An example of a graph satisfying the hypotheses of the theorem is the graph of the dodecahedron. This has $f = f_5 = 12$, and $g = 11$.

(d) It is not hard to see that there is no planar cubic graph G satisfying the hypotheses of the theorem with $g = 12$. However, I conjecture that for sufficiently large g, such graphs should exist (and be rather plentiful).

The author would like to thank the organizers of the 1988 Algebraic Geometry Conference held in Sundance, Utah, for their kind invitation and for providing a pleasant and extremely stimulating week of mathematics. I would also like to thank the Departments of Mathematics at the Universities of Pisa and Rome II where this work was completed. I want to especially express my appreciation to Prof. C. Ciliberto for many useful conversations on this topic.

2. Global sections of ω_X for a planar graph curve

By Remark $(1.3)(a)$, we have that G has no loops or multiple edges; therefore the associated graph curve $X = X_G$ has all its components smooth rational curves, any two meeting at most once, in an ordinary node. Moreover, each component has exactly three nodes on it, since G is a cubic graph. Any global section σ of ω_X restricts to a meromorphic 1-form σ_v on each component C_v of X, with at most one simple pole at each node; by the residue theorem, the sum of the three residues of σ_v at these three nodes is zero. These meromorphic 1-forms "fit together" in the following sense: if P_e is the node where components C_v and C_w meet then the residue of σ_v at P_e and the residue of σ_w at P_e sum to zero.

Given a global section σ of ω_X, a vertex v of G, and an edge e containing v, denote by $\text{res}(\sigma, v, e)$ the residue of the meromorphic 1-form σ_v on C_v at the point P_e. In this way to each global section σ of ω_X we associate $6g-6$ complex numbers, namely the set $\{\text{res}(\sigma, v, e)|v \text{ is on } e\}$; this assignment maps $H^0(X, \omega_X)$ into $\mathbb{C}^{6g-6}$. Note that "res" is linear in the first argument.

The two sets of conditions mentioned above, namely

(2.1) $$\text{res}(\sigma, v, e_1) + \text{res}(\sigma, v, e_2) + \text{res}(\sigma, v, e_3) = 0,$$

if $\{e_1, e_2, e_3\}$ are the three edges containing v,

(2.2) $$\text{res}(\sigma, v_1, e) + \text{res}(\sigma, v_2, e) = 0,$$

if $\{v_1, v_2\}$ are the two vertices of e,

impose a total of $5g-6$ independent conditions on the $6g-6$ numbers: there are $v+e = 5g-5$ conditions, but they are not independent. Indeed, the sum of all $6g-6$ numbers is zero, and this follows from either (2.1) or (2.2). Hence the assignment of these residue numbers to global sections of ω_X embeds $H^0(X, \omega_X)$ isomorphically onto the subspace of $\mathbb{C}^{6g-6}$ determined by these conditions. I.e.,

2.3 PROPOSITION. *Given a set of $6g-6$ complex numbers indexed by pairs (v, e) with v on e, satisfying conditions (2.1) and (2.2), there is a unique element of $H^0(X, \omega_X)$ with this given set of residue numbers.*

This is discussed in §3 of [C-H-M], and is independent of the planarity of G. By using this planarity, we can obtain a natural spanning set of $g + 1$ sections of ω_X, corresponding to the $g + 1$ faces of the decomposition of the sphere. For this we choose once and for all an orientation of the sphere.

Let f be a face of the decomposition. By the previous proposition, to define a global section of ω_X, we must give a complex number to every adjacent vertex-edge pair, subject to the conditions that the sum of the three numbers given at a vertex is 0, and the sum of the two number given at an edge is 0. Therefore define a section σ_f corresponding to the face f by declaring

(2.4) $$\text{res}(\sigma_f, v, e) = 0, \quad \text{if } v \text{ is not in the face of } f,$$

and if v is on f, with its three edges being e_1, e_2, e_3, set

$$\text{res}(\sigma_f, v, e_1) = 0, \quad \text{res}(\sigma_f, v, e_2) = 1, \quad \text{and} \quad \text{res}(\sigma_f, v, e_3) = -1,$$

where e_1 is the edge not adjacent to f, and the edges e_1, e_2, e_3 are in counterclockwise order around the vertex v (Figure 3):

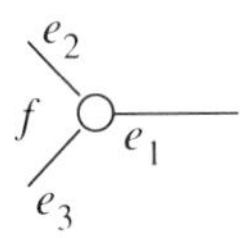

FIGURE 3

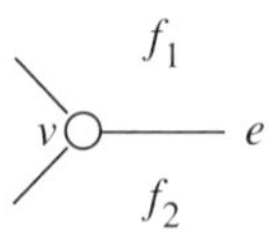

FIGURE 4

It is immediate that the assignments of (2.4) satisfy the conditions (2.1) and (2.2), and so define a section σ_f of ω_X.

2.5 LEMMA. *The sections* $\{\sigma_f | f \text{ is a face}\}$ *span* $H^0(X, \omega_X)$. *In particular, the only linear relation among the* σ_f's *is that their sum is zero.*

PROOF. Since $H^0(X, \omega_X)$ has dimension g, and there are $g + 1$ faces, the last statement implies the first. First let us check that $\sum \sigma_f = 0$. Fix a vertex v and an edge e containing v. Then

$$\text{res}(\sum \sigma_f, v, e) = \Sigma \text{res}(\sigma_f, v, e) = \text{res}(\sigma_{f_1}, v, e) + \text{res}(\sigma_{f_2}, v, e),$$

where f_1 and f_2 are the two faces adjacent to e, since for all other faces f, $\text{res}(\sigma_f, v, e) = 0$. We may assume that the graph locally near v is the one in Figure 4. Thus, $\text{res}(\sigma_{f_1}, v, e) = +1$ and $\text{res}(\sigma_{f_2}, v, e) = -1$ by (2.4); therefore $\text{res}(\sum \sigma_f, v, e) = 0$ for all pairs (v, e). Hence $\sum \sigma_f = 0$ as claimed.

Now assume that $\sum c_f \sigma_f = 0$ for complex numbers c_f, not all zero. We need to show that all c_f are the same, to finish the proof of the Lemma.

Again fix a vertex v on an edge e; since $\sum c_f \sigma_f = 0$, we must have $\text{res}(\sum c_f, \sigma_f, v, e) = 0$. If as above f_1 and f_2 are the two faces adjacent to e, this implies that $0 = c_{f_1} \text{res}(\sigma_{f_1}, v, e) + c_{f_2} \text{res}(\sigma_{f_2}, v, e) = c_{f_1} - c_{f_2}$. This shows that whenever f_1 and f_2 are faces with an edge e in common, then $c_{f_1} = c_{f_2}$; since the dual graph of the decomposition of the sphere given by G is connected, this shows that all c_f are equal and finishes the proof of the Lemma. $\square$

3. Oriented coordinates and the target of the Gaussian map

As in the previous section, we assume that an orientation of the sphere has been fixed. Let v be a vertex of G representing a component $C_v \cong \mathbb{P}^1$ of $X = X_G$. An *oriented coordinate* on C_v is an affine coordinate z on C_v such that at the three nodes of X on C_v, z take on the values $0, 1, \infty$ in a counterclockwise order, considering the nodes as corresponding to the edges of G emanating from v.

On every component of X there are then three possible oriented coordinates: if z is one, then the other two are $(z - 1)/z$ and $1/(1 - z)$. Note that at each node of C_v, there is a unique oriented coordinate on C_v which is zero at that node.

Choose once and for all an oriented coordinate z_v on every component C_v of X.

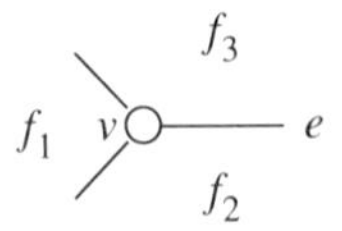

FIGURE 5

We will use as a local coordinate at the node P_e of C_v corresponding to the edge e of G containing v the unique oriented coordinate $t_{v,e}$ on C_v which is zero at P_e. In other words,

(3.1)   we use $t_{v,e} = z_v$ if z_v is zero at P_e,

we use $t_{v,e} = (z_v - 1)/z_v$ if z_v is 1 at P_e, and

we use $t_{v,e} = 1/(1 - z_v)$ if z_v is ∞ at P_e.

This then defines a global coordinate z_v for each component C_v of X, and a local coordinate $t_{v,e}$ at each node P_e of C_v, which is zero at the node.

It will be useful to write the spanning forms σ_f in terms of these local coordinates on each component. Fix a component C_v and a node P_e on C_v. Of course $\sigma_{f|_{C_v}} = 0$ if the vertex v is not on the face f. In case v is on f, we have three possibilities, according to the position of e relative to f. Locally near v, we have the picture in Figure 5, and f could be in the position of f_1, f_2, or f_3. An easy calculation using the definition (2.4) for σ_f gives the following formula for σ_f in terms of the local oriented coordinate $t_{v,e}$ on C_v at P_e.

$$(3.2) \qquad \sigma_{f|_{C_v}} = \begin{cases} dt_{v,e}/(t_{v,e} - 1) & \text{if } f = f_1 \text{ above,} \\ -dt_{v,e}/t_{v,e} & \text{if } f = f_2 \text{ above,} \\ -dt_{v,e}/(t_{v,e}(t_{v,e} - 1)) & \text{if } f = f_3 \text{ above.} \end{cases}$$

Let us now briefly recall how the choices of these local coordinates gives an isomorphism of the target $H^0(X, \Omega^1_X \otimes \omega^2_X)$ with $\mathbb{C}^{5g-5}$, as explained in [C-H-M, §4].

On the graph curve X, Ω^1_X is not locally free; the torsion part of Ω^1_X is supported at the $3g - 3$ nodes of X, and at each node this torsion part is a 1-dimensional skyscraper sheaf. Modulo torsion, Ω^1_X is the direct sum over the components C_v of X of the sheaves $\Omega^1_{C_v}$. In other words, one has a short exact sequence

$$(3.3) \qquad 0 \to \text{Tors}(\Omega^1_X) \to \Omega^1_X \to \bigoplus_v \Omega^1_{C_v} \to 0,$$

where to be perfectly precise the direct sum on the right is the direct image of Ω^1 of the normalization of X, which is of course the disjoint union of the components C_v. After tensoring with the invertible sheaf ω^2_X and taking global sections, we have the sequence

$$(3.4) \quad 0 \to H^0(\text{Tors}(\Omega^1_X) \otimes \omega^2_X) \to H^0(\Omega^1_X \otimes \omega^2_X) \to \bigoplus_v H^0(\omega^2_X \otimes \Omega^1_{C_v}) \to 0,$$

which is exact since, as remarked above, $\text{Tors}(\Omega_X^1)$ is supported at the nodes, and therefore the H^1 vanishes.

For any component C_v of X, $\omega_{X|_{C_v}}$ is the sheaf of meromorphic 1-forms on C_v with at most simple poles at the three nodes of C_v; hence $\omega_{X|_{C_v}}$ is a line bundle of degree one on C_v. Therefore $\omega_X^2 \otimes \Omega_{C_v}^1$ is a line bundle on C_v of degree 0, and is therefore trivial, since C_v is isomorphic to $\mathbb{P}^1$. Moreover, with our choice of affine coordinate z_v on C_v, we have that

$$(3.5) \qquad \alpha_v = dz_v^{\otimes 3}/(z_v(z_v - 1))^2,$$

is a basis vector for $H^0(\omega_X^2 \otimes \Omega_{C_v}^1)$.

At a node P_e of X, let t_1 and t_2 be the two local coordinates on the two components through P_e at P_e, as defined in (3.1). Then the dualizing sheaf ω_X is generated at P_e by the form $dt_1/t_1 - dt_2/t_2$, and the Kähler differentials Ω_X^1 are generated at P_e by dt_1 and dt_2, subject to the relation $t_2\, dt_1 + t_1\, dt_2 = 0$. Therefore at P_e, the torsion in $\omega_X^2 \otimes \Omega_X^1$ is 1-dimensional at P_e, and is generated by

$$(3.6) \qquad \beta_e = (dt_1/t_1 - dt_2/t_2)^2 \cdot (t_2\, dt_1 - t_1\, dt_2).$$

Since we have chosen the local coordinates at each node, this generator is well-defined up to sign: the sign changes if we switch the roles of t_1 and t_2.

Note that the forms α_v can be considered as global sections of $\omega_X^2 \otimes \Omega_X^1$, by extending them to zero on all components other than C_v. This then gives a splitting of the sequence (3.4), and we obtain an explicit isomorphism

$$(3.7)$$
$$H^0(\Omega_X^1 \otimes \omega_X^2) \cong H^0(\text{Tors}(\Omega_X^1) \otimes \omega_X^2) \oplus \bigoplus_v H^0(\omega_X^2 \otimes \Omega_{C_v}^1) \cong \mathbb{C}^{3g-3} \oplus \mathbb{C}^{2g-2}$$

$$\cong \mathbb{C}^{5g-5},$$

as claimed above. We will refer to the composition of the Gaussian map ϕ with these $5g - 5$ projections by ϕ_e and ϕ_v; therefore, to determine the corank of ϕ it is enough to determine the corank of $\bigoplus_e \phi_e \oplus \bigoplus_v \phi_v$.

4. The proof of the main theorem

We are now in a position to prove Theorem 1.2. As in the previous sections, choose an orientation for the sphere, and choose oriented coordinates z_v for each component C_v of $X = X_G$, which induce local coordinates $t_{v,e}$ at each node P_e as defined in (3.1).

Since the collection $\{\sigma_f | f \text{ is a face}\}$ spans $H^0(\omega_X)$, the elements $\sigma_{f_1} \wedge \sigma_{f_2}$ span the domain, $\wedge^2 H^0(\omega_X)$, of the Gaussian map ϕ. Therefore the elements $\phi(\sigma_{f_1} \wedge \sigma_{f_2})$ (where f_1 and f_2 are distinct faces) span the image of

ϕ in $H^0(\Omega^1_X \otimes \omega^2_X)$. By the assumptions of the theorem, given two distinct faces f_1 and f_2, we have the following three possibilities:

    (1) f_1 and f_2 have disjoint edge-neighborhoods,

    (2) f_1 and f_2 have a single edge e in their common edge-neighborhoods,

    (3) f_1 and f_2 are the faces adjacent to an edge e, and have only e and the two vertices of e in their common closure.

In case (1), the forms σ_{f_1} and σ_{f_2} have disjoint support, so that $\sigma_{f_1} \wedge \sigma_{f_2}$ is zero; hence these do not contribute to the image of ϕ.

In case (2), the forms σ_{f_1} and σ_{f_2} have only P_e in their common support. Therefore $\phi_v(\sigma_{f_1} \wedge \sigma_{f_2}) = 0$ for all v, and $\phi_{e'}(\sigma_{f_1} \wedge \sigma_{f_2}) = 0$ for all edges $e' \neq e$. A calculation is necessary to determine $\phi_e(\sigma_{f_1} \wedge \sigma_{f_2})$.

4.1. LEMMA. *Assume that f_1 and f_2 have a single edge e in their common edge-neighborhoods. Then $\phi_e(\sigma_{f_1} \wedge \sigma_{f_2}) \neq 0$.*

PROOF. Let v_1 and v_2 be the two vertices on the edge e, such that v_1 is on f_1 and v_2 is on f_2. We have the picture near e given in Figure 6.

Let C_1 and C_2 be the two components of X corresponding to v_1 and v_2 respectively. Hence P_e is the intersection of C_1 and C_2, and we have the chosen oriented coordinates $t_1 = t_{v_1,e}$ on C_1 at P_e and $t_2 = t_{v_2,e}$ on C_2 at P_e: note that $t_1 t_2 = 0$ near P_e. According to (3.2), we have

$$\sigma_{f_1} = dt_1/(t_1 - 1) \quad \text{and} \quad \sigma_{f_2} = dt_2/(t_2 - 1),$$

near P_e. Unfortunately these are not written in terms of the local generator $\omega = dt_1/t_1 - dt_2/t_2$ for ω_X at P_e. To do so, note that since $t_1 dt_2 + t_2 dt_1 = 0$, we have $dt_2/t_2 + dt_1/t_1 = 0$; hence $dt_1/t_1 = \frac{1}{2}\omega$, and similarly $dt_2/t_2 = -\frac{1}{2}\omega$. Therefore,

$$\sigma_{f_1} = [\tfrac{1}{2}t_1/(t_1 - 1)]\omega \quad \text{and} \quad \sigma_{f_2} = [-\tfrac{1}{2}t_2/(t_2 - 1)]\omega,$$

and we are now in a position to calculate $\phi_e(\sigma_{f_1} \wedge \sigma_{f_2})$:

$$\phi(\sigma_{f_1} \wedge \sigma_{f_2}) = [[\tfrac{1}{2}t_1/(t_1-1)]\,d[-\tfrac{1}{2}t_2/(t_2-1)] - [-\tfrac{1}{2}t_2/(t_2-1)]\,d[\tfrac{1}{2}t_1/(t_1-1)]]\omega^2$$

$$= \frac{1}{4}[[t_1/(t_1-1)][dt_2/(t_2-1)^2] - [t_2/(t_2-1)][dt_1/(t_1-1)^2]]\omega^2$$

$$= \frac{1}{4}[t_2\,dt_1 - t_1\,dt_2 + t_1^2\,dt_2 - t_2^2\,dt_1]\omega^2/(t_1-1)^2(t_2-1)^2$$

$$= \frac{1}{4(t_1-1)^2(t_2-1)^2} \cdot (t_2\,dt_1 - t_1\,dt_2)\omega^2 = \frac{1}{4(t_1-1)^2(t_2-1)^2} \cdot \beta_e.$$

Note that we have used that $t_1^2\,dt_2 = t_1(t_1\,dt_2) = t_1(-t_2\,dt_1) = 0$, and similarly $t_2^2\,dt_1 = 0$, since $t_1 t_2 = 0$.

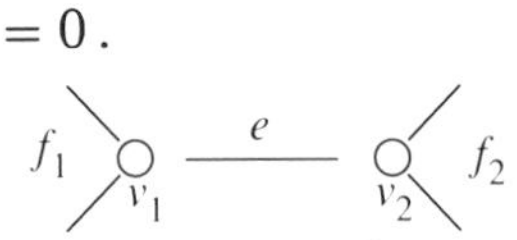

FIGURE 6

Recall that β_e is the local generator at P_e for the torsion in $\Omega^1_X \otimes \omega^2_X$, and to finish the proof we note that $\phi_e(\sigma_{f_1} \wedge \sigma_{f_2})$ is obtained from the above formula simply by setting t_1 and t_2 equal to zero, and looking at the coefficient of β_e: thus we obtain $\phi_e(\sigma_{f_1} \wedge \sigma_{f_2}) = 1/4$, which is nonzero. $\square$

From the above computation, we see that the torsion part of $H^0(X, \Omega^1_X \otimes \omega^2_X)$ is in the image of the Gaussian map ϕ: to hit a generator of the torsion part at a node P_e, use $4\sigma_{f_1} \wedge \sigma_{f_2}$, where f_1 and f_2 are the two faces neighboring the edge e. Therefore, the corank of ϕ is equal to the corank of the composition of ϕ with the projection onto the global sections of $\Omega^1_X \otimes \omega^2_X$ modulo torsion, and as remarked in the previous section, this has been identified with $\mathbb{C}^{2g-2}$, with one generator α_v (defined in (3.5)) corresponding to each vertex v of G. To finish the proof of the theorem, it now suffices to show that this composition has corank one.

We have left to use only the elements $\sigma_{f_1} \wedge \sigma_{f_2}$ where f_1 and f_2 satisfy condition (3) above, namely that they share an edge e. For each edge e of the graph G there is a unique (up to sign) such element of $\wedge^2 H^0(\omega_X)$, which we will call $\tau_e : \tau_e = \sigma_{f_1} \wedge \sigma_{f_2}$, where f_1 and f_2 are the two faces adjacent to e. Therefore the corank of ϕ is the same as the corank of the map $\Phi : \bigoplus \mathbb{C} \cdot \tau_e \to \bigoplus_v \mathbb{C} \cdot \alpha_v$, which is the composition of the Gaussian map ϕ on the basis $\{\tau_e\}$ of the domain, with the projection onto the global sections of $\Omega^1_X \otimes \omega^2_X$ modulo torsion.

We must now turn to computing $\Phi(\tau_e)$; we will denote by $\Phi_v(\tau_e)$ the coefficient of α_v in $\Phi(\tau_e)$. Note that $\Phi_v(\tau_e) = 0$ unless v is one of the two vertices on the edge e.

4.2. LEMMA. *Let v_1 and v_2 be the two vertices on the edge e. Then* $\Phi_{v_1}(\tau_e) = -\Phi_{v_2}(\tau_e) = \pm 1$.

PROOF. Of course the ambiguity in the sign of these quantities is due to the ambiguity of τ_e. The content of the Lemma is that the values of $\Phi(\tau_e)$ are nonzero and opposite in the two coordinates indexed by v_1 and v_2.

The proof is simply a straightforward calculation in the spirit of the proof of Lemma 4.1. The proof is complicated by the fact that the generator α_v is defined using one of the oriented coordinates on C_v, and there are three such; hence a priori we must make three computations to verify the result.

However it is a somewhat remarkable fact that the generator α_v does not depend on the choice of a local oriented coordinate. To check this, suppose z_v is the chosen oriented coordinate, so that $x = (z_v - 1)/z_v$ and $y = 1/(1 - z_v)$ are the two others, by (3.1). Then a simple computation shows that $(dz_v)^3/(z_v(z_v - 1))^2 = (dx)^3/(x(x - 1))^2 = (dy)^3/(y(y - 1))^2$, which shows the independence of α_v on the choice of local oriented coordinate.

Therefore to compute $\Phi_v(\tau_e)$, we may use any of the three local oriented coordinates on C_v. It is convenient to use $t_{v,e}$, the oriented coordinate which is zero at P_e; this is what we will do.

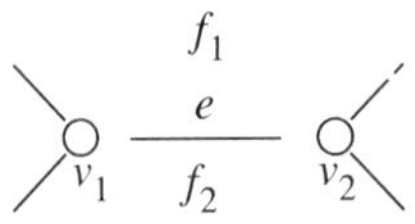

FIGURE 7

The picture of the graph near the edge e is given in Figure 7, and in terms of the local coordinate $t = t_{v_1, e}$ on C_{v_1} at P_e, we have

$$\sigma_{f_1} = -dt/(t(t-1)) \quad \text{and} \quad \sigma_{f_2} = -dt/t,$$

on C_{v_1}, by (3.2). Therefore locally, using the parameter t, we have

$$\phi(\tau_e) = \phi(-dt/(t(t-1)) \wedge -dt/t) = \phi(dt/(t(t-1)) \wedge dt/t)$$
$$= [[1/(t(t-1))]\, d[1/t] - [1-t]\, d[1/(t(t-1))]](dt)^2$$
$$= (dt)^3/(t(t-1))^2 = \alpha_{v_1},$$

so that $\Phi_{v_1}(\tau_e) = +1$. Now by symmetry we see that $\Phi_{v_2}(\tau_e) = -1$; replacing v_1 by v_2 in the above computation has the effect of switching the roles of f_1 and f_2, which just changes the sign of τ_e. $\square$

This is the final ingredient. Upon choosing directions for the edges of G, whish essentially gives a choice of the sign of τ_e for each e, we see that the map $\Phi : \bigoplus_e \mathbb{C} \cdot \tau_e \to \bigoplus_v \mathbb{C} \cdot \alpha_v$ is the same as the boundary map on the chains of G, from the 1-chains to the 0-chains, considering G as a one-dimensional simplicial complex. Therefore the cokernel of this map Φ is isomorphic to $H^0(G, \mathbb{C})$, where now G is considered as a topological space. Since G is connected, we have that this space has dimension one, and the theorem is proved.

REFERENCES

[C-H-M] C. Ciliberto, J. Harris, and R. Miranda, *On the surjectivity of the Wahl map*, Duke Math. J. **57** (1988), 829–858.

[C-M1] C. Ciliberto and R. Miranda, *Gaussian maps for certain families of canonical curves*, Duke Math. J. (to appear).

[C-M2] ____, *On the Gaussian map for canonical curves of low genus*, Duke Math. J. **61** (1990), 417–443.

[W] J. Wahl, *The Jacobian algebra of a graded Gorenstein singularity*, Duke Math. J. **55** (1987), 843–871.

DEPARTMENT OF MATHEMATICS, COLORADO STATE UNIVERSITY, FORT COLLINS, CO 80523

Absence of the Veronese from Smooth Threefolds in $\mathbb{P}^5$

ZIV RAN

The purpose of this note is to verify the statement of the title, thus answering a question raised by Sheldon Katz during the conference (and earlier, he tells me). We will prove the following.

THEOREM. *Let* S *denote the Veronese surface in* $\mathbb{P}^5$*, image of the "2-uple" embedding*

$$\varphi = \varphi_{|\mathscr{O}(2)|} : \mathbb{P}^2 \to \mathbb{P}^5 .$$

Then any threefold $T \subset \mathbb{P}^5$ *containing* S *and smooth along it must be a cone.*

PROOF. We begin by recalling a couple of well-known properties of S, which may be found in Semple-Roth or Griffiths-Harris, among other places.

1. The secant lines of S fill up a cubic hypersurface F, which has an A_1-singularity along S.

2. S has no trisecant lines; i.e., no line of $\mathbb{P}^5$ meets S in 3 or more points.

(For (2) it suffices to note that any length-3 subscheme of $\mathbb{P}^2$ imposes independent conditions on conics.)

We claim that the threefold T cannot be contained in F: indeed if it were, it would give rise to a cross-section of the projectivized normal cone to S in F; however, the latter is a conic bundle which may easily be described explicitly and shown to be sectionless.

Actually, the foregoing argument shows a bit more: namely that at a general point of S the (embedded) tangent space of T cannot be contained in the tangent cone of F, hence we have an equality of divisors on T:

$$F \cdot T = 2S + R ,$$

1980 *Mathematics Subject Classification* (1985 *Revision*). Primary 14J25, 14E25.

This research was partially supported by the NSF grant DMS 86-12391 and by the Sloan foundation.

This paper is in final form and no version of it will be submitted for publication elsewhere.

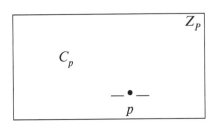

FIGURE 1

where $R \not\supseteq S$. Note that $R \neq \varnothing$ (because, if for no other reason, we have $3 = \deg F \dagger 4 = \deg S$).

Now for any point $p \in S$, the embedded tangent space $\tilde{T}_p T$ meets S in a scheme which corresponds via φ to the intersection of 2 conics in $\mathbb{P}^2$, singular at $\varphi^{-1}(p)$, hence we have either

(a) $\tilde{T}_p T \cap S$ is a complete intersection scheme of length 4 supported at p; or

(b) $(\tilde{T}_p T \cap S)_{\mathrm{red}} = C_p$ is a smooth conic (i.e., $\varphi^{-1}(C_p)$ is a line).

Now set $D = S \cap R$. As $F \cdot T$ is ample on T, note that D is a nonempty curve. We claim that for $p \in D$, case (b) above must occur. To see this, start with a general point $q \in R$, and note by (1) that q lies on ∞^1 secants $\overline{p'p''}$ of S, where the points $p', p'' \in S$ fill up a curve (in fact, as is easy to see, a smooth conic). Now specialize q to p. Then p' and p'' cannot both specialize to p, since they must still fill up a curve. If neither p' nor p'' specialized to p, we would get a trisecant line of S, which cannot exist. Thus we may assume both p' and q, but not p'', specialize to p, hence the limiting position of $\overline{p'p''}$ is a tangent line to T at p meeting S additionally at $p'' \neq p$, so that we are in case (b). (Actually, the converse is also true: if $p \in S \backslash D$, then case (a) must occur; but we will not need this.)

Thus we get for each $p \in D$ a smooth conic $C_p \subset S$ whose span Z_p is a $\mathbb{P}^2$ contained in $\tilde{T}_p T$ as shown in Figure 1.

Putting these together, we get a $\mathbb{P}^2$-bundle

$$Z$$

$$D$$

and we may write $Z = \mathbb{P}(E)$ where E is a rank-3 subbundle of the trivial bundle $V \otimes \mathscr{O}_D$, where $\mathbb{P}^5 = \mathbb{P}(V)$. Note that E carries a natural filtration

$$E = E_3 \supset E_2 \supset E_1,$$

where E_1 is the tautological subline bundle and E_2 is the rank-2 subbundle such that for each $p \in D$, we have $\mathbb{P}(E_2(p)) = \tilde{T}_p C_p$.

Now write $\tilde{\eta}$ for the positive generator of $\mathrm{Pic}\, S$ and η for its restriction on D. Then we have

$$E_1 = \mathscr{O}_D(-1) = \eta^{-2}.$$

As $Z_p \cap \tilde{T}_p S = \tilde{T}_p C$, it follows that

$$E_3/E_2 \simeq N_{S/T}(-1) \otimes \mathcal{O}_D,$$

hence, writing $N_{S/T} = \tilde{\eta}^{-s}$, we have

$$E_3/E_2 \simeq \eta^{-s-2}.$$

To compute E_2/E_1, note the commutative diagram

$$
\begin{array}{ccc}
N_{T/\mathbb{P}^5}(-1)^{\vee} \otimes \mathcal{O}_D & \xrightarrow{\psi} & \mathrm{Sym}^2(T_S \otimes \eta^{-1})^{\vee} \\
\cap & & \cap \\
V^{\vee} \otimes \mathcal{O}_D & \xrightarrow{\sim} & \mathrm{Sym}^2(W \otimes \mathcal{O}_D)^{\vee}
\end{array}
$$

where $S = \mathbb{P}(W)$, and we have

$$
\begin{aligned}
\mathrm{im}\,\psi &\simeq (T_S \otimes \eta^{-1})^{\vee} \otimes ((E_2/E_1) \otimes \eta)^{\perp} \\
&\simeq (T_S(-1))^{\vee} \otimes ((E_2/E_1))^{\perp} \otimes \mathcal{O}_D(-1) \\
&\simeq (T_S(-1))^{\vee} \otimes (T_S(-1)/(E_2/E_1))^{\vee} \otimes \mathcal{O}_D(-1).
\end{aligned}
$$

Now by an easy calculation we have

$$c_1(N_{T/\mathbb{P}^5}(-1) \otimes \mathcal{O}_D) = \eta^{s+1}, \qquad c_1(T_T(-1) \otimes \mathcal{O}_D) = \eta^{-s-3}.$$

Putting $A = T_S(-1) \otimes \mathcal{O}_D/(E_2/E_1)$, we conclude

$$(1) \qquad \eta^{s+1} = A^{-1} \otimes \eta^{-1}.$$

On the other hand, we have

$$T_T(-1)/E_3 \simeq A,$$

hence, equating c_1 we get

$$
(2) \qquad
\begin{aligned}
\eta^{s-3} &= A \otimes E_3/E_2 \otimes E_2/E_1 \\
&= A \otimes E_2/E_1 \otimes \eta^{-s-2}.
\end{aligned}
$$

Comparing (1) with (2) we conclude

$$E_2/E_1 = \eta^{s+1}.$$

Now we can easily complete the proof. The fact that our $\mathbb{P}^2$-bundle Z carries a bundle of smooth conics yields an isomorphism

$$\alpha : E \to E^{\vee} \otimes M,$$

for some line bundle M on D. Moreover, considering the way our filtration $E_{\bullet}$ was defined, we see easily that α must send $E_{\bullet}$ to the dual filtration on $E^{\vee} \otimes M$. By considering the gradeds of the two filtrations, we see that $M = \eta^{-s-4} = \eta^{2s+2}$.

Since D is a nonempty curve, it follows that $s = -2$; i.e., $N_{S/T} = \eta^2 = \mathcal{O}_S(1)$. But from the normal bundle sequence on S it follows easily, using

Bott vanishing, that any subbundle of $N_{S/\mathbb{P}^5}$ isomorphic to $\mathscr{O}_S(1)$ must come, via the map

$$V \otimes \mathscr{O}_S(1) \to T_{\mathbb{P}^5} \otimes \mathscr{O}_S \to N_{S/\mathbb{P}^5},$$

from a unique point $q = q(S) \in \mathbb{P}^5 = \mathbb{P}(V)$; in our case q is characterized by the fact that all the embedded tangent spaces $\tilde{T}_p T$ for $p \in S$ contain q.

Now note finally that, since $h^0(N_{S/T}) = 6$, $h^1(N_{S/T}) = 0$, S must move on T in an effective 6-parameter family $\{S_b : b \in B\}$, where, of course, almost all the S_b are projectively equivalent to S and by the above, we may map $B \xrightarrow{q} \mathbb{P}^5$ by sending $b \in B$ to the point $q(S_b) \in \mathbb{P}^5$. Considering a general fibre $q^{-1}(q_0)$ of q, we find a family of S_b's filling up T with fixed $q = q_0$, and conclude that almost all embedded tangent spaces $\tilde{T}_p T$, $p \in T$, contain q_0, so that T is a cone. $\square$

Note incidentally that conversely, a general cone in $\mathbb{P}^5$ over the Veronese S is indeed smooth along S (equivalently, a general projection of S to $\mathbb{P}^4$ is smooth, cf. Property 1 above), so those exceptions to the Theorem do, in fact, occur.

REMARK. Presumably, the motivation behind Katz's question was that its solution might have shed some light on the more general question of what surfaces may lie on a smooth threefold in $\mathbb{P}^5$. Unfortunately though, the foregoing argument, being so closely tied up with the very special properties of the Veronese, seems to shed no such light. To illustrate my ignorance, the Veronese may, for all I know, be the *only* smooth surface in $\mathbb{P}^5$ not lying on a smooth threefold!

ACKNOWLEDGMENT. I am grateful to Sheldon Katz for raising this cute question during the conference (more precisely, I heard it on the limo ride back to the airport, but this is morally the same), and to David Eisenbud, Joe Harris, and especially Bob Speiser for organizing the conference and helping make it such a pleasant experience.

UNIVERSITY OF CALIFORNIA, RIVERSIDE, CA 92521

Contemporary Mathematics
Volume **116**, 1991

Limits of Conormal Schemes

ROBERT SPEISER

This article began as a lecture at the 1988 Sundance conference, about Van Gastel's recent and useful description [**VG**] of the degenerations of the conormal schemes of hypersurfaces in characteristic 0, using limits of refined polar classes. Conversations at Sundance with Steve Kleiman led to a new and simpler approach, in any characteristic, which became the subject of the lecture. After the conference, further insight quickly yielded generalizations to local complete intersections of any codimension; we describe these too.

Polar varieties and their classes date back to the 19th century. Recent work provides a fully modern impetus. In particular, results by Lê and Teissier ([**LT**] and [**T**]), by Kleiman [**KO**], by Piene [**P**] and by Henry, Merle and Sabbah [**HMS**] provide groundwork for Van Gastel's constructions.

The approach we follow, while inspired in some respects by Van Gastel's work, takes a different point of departure: the straightforward treatment of polar classes of hypersurfaces given by Fulton [**F**, pp. 84-5], in the context of the intersection theory he developed jointly with MacPherson. For hypersurfaces, the main idea is to blow up the Jacobian locus, as in [**P**], compute in the Picard group of the blowup, and then push down. This gives, for example, the standard expression of the polar classes in terms of the hyperplane class and the Segre classes of the Jacobian. Special cases include Plücker's formula for the class of a hypersurface, and the well-known formula for the degree of the dual variety.

To study a conormal family, however, pushing down to the base to compute polars seems odd, because we need to go up to the conormal scheme instead! Nonetheless, the pushdown works in characteristic 0, because there a limit conormal scheme is completely determined by information on the base, and so Van Gastel obtains his results. For the general case, we must explore the interplay between the base family and the conormal family more carefully.

1980 *Mathematics Subject Classification* (1985 *Revision*). Primary 14C17, 14N05.
The author was partly supported by NSF grant DMS-8802015.
This paper is in final form and no version of it will be submitted for publication elsewhere.

In this paper, we generalize Van Gastel's work to local complete intersections of arbitrary codimension, eventually with isolated singularities, in any characteristic. In short, our strategy is to understand the conormal family through its relation to the deformation space. More precisely, to bridge between the Jacobian locus on X and the conormal scheme, we use the functor T^1 of Lichtenbaum and Schlessinger [LS]. We define $\tilde{X}$ to be the blowup of the projective conormal bundle along the support of $T^1(\mathscr{O}_X)$. Then $\tilde{X}$ dominates X, and the exceptional divisor of $\tilde{X}$ projects onto the Jacobian locus of X; in this setting it is natural to give the Jacobian a scheme structure via the annihilator of $T^1(\mathscr{O}_X)$. For a hypersurface, $\tilde{X}$ is precisely the conormal scheme; in general, $\tilde{X}$ dominates the conormal scheme birationally. To generalize the calculation in [F], we define *superpolar* schemes and classes on $\tilde{X}$. The superpolar classes are expressed in terms of *superSegre* classes of the Jacobian locus; these, too, live on $\tilde{X}$. The polar cycles on the base, defined in this approach as pushdowns from $\tilde{X}$, reflect geometry above.

To simplify the exposition, the paper is divided into two parts, proceeding from the special to the general. The first part is about hypersurfaces, and is more expository, concentrating on Van Gastel's results about hypersurfaces. The second part treats arbitrary local complete intersections. Here the exposition gives more details because the work to be described is new.

Superpolars are introduced, in §1, for hypersurfaces, to be generalized later. In §2, we indicate a new proof of Van Gastel's description of the conormal scheme of a limit hypersurface with isolated singularities. This proof is joint work, done at Sundance, with Steve Kleiman. §3 completes the Sundance lecture and the first part of this paper. There we sketch Van Gastel's second main result, a description of the conormal scheme of a possibly nonreduced limit of plane curves, when the Kodaira-Spencer section is not identically zero. Because important cases are not covered by Van Gastel's analysis, we give some examples, and indicate a different method for treating limit plane curves, inspired by Zeuthen and modernized in [KS].

In the rest of the paper, we consider local complete intersections of any codimension, in any characteristic. After the preliminary §4, we define the superpolar subschemes and their classes on $\tilde{X}$ in §5, and obtain formulas the polar classes as refined direct images on X, by straightforward calculation, in §6. Finally, in §7, we generalize Van Gastel's limit theorem from hypersurfaces (§2) to local complete intersections with isolated singularities.

In addition to Steve Kleiman, who collaborated on the first two sections, I would like to thank Lawrence Ein, Bill Lang, Ragni Piene and the referee, for encouragement and help along the way.

1. Hypersurfaces

Suppose given a hypersurface $X \subset \mathbf{P}^{n+1}$ of degree d; we have

$$X = V(F(X_0, \ldots, X_{n+1})),$$

where $F = F(X_0, \ldots, X_{n+1})$ is a homogeneous polynomial of degree d in a given system of homogeneous coordinates. Denote by J the Jacobian subscheme of X. We have

$$J = V(\partial F / \partial X_0, \ldots, \partial F / \partial X_{n+1}).$$

Write $\tilde{X}$ for the blowup of X along J, and write $\check{\mathbf{P}}^{n+1}$ for the dual projective space. We have a diagram

where $f = (\partial f / \partial X_0, \ldots, \partial F / \partial X_{n+1})$ is induced by the the Gauss map $\gamma : X_{\mathrm{reg}} \to \check{\mathbf{P}}^{n+1}$; by definition, for each smooth point $x \in X$, viewed as a point of $\tilde{X}$, we have $f(x) = T_x X$, the tangent hyperplane. Hence $\tilde{X}$ is the closure of the graph of γ in $\check{\mathbf{P}}^{n+1}$, (as observed in [P], for example), by definition [KC] the *conormal scheme* of $X \subset \mathbf{P}^{n+1}$. Clearly $\tilde{X}$ embeds in the incidence correspondence

$$I = \left\{ (p, H) \in \mathbf{P}^{n+1} \times \check{\mathbf{P}}^{n+1} | p \in H \right\}.$$

Fix a linear subspace $W_k \subset \mathbf{P}^{n+1}$ of dimension $k - 1$. The dual subspace

$$W^\vee = \{ H \in \check{\mathbf{P}}^{n+1} | H \supset W \},$$

has codimension k in $\check{\mathbf{P}}^{n+1}$. Set

$$M_k = M_k(X, W) = \text{closure in } X \text{ of } \{ p \in X_{\mathrm{reg}} | T_p X \supset W_k \},$$

where X_{reg} denotes the set of regular points of X. As a point set, $M_k = \pi(f^{-1}(W^\vee))$, by continuity and the definition of $\tilde{X}$. For general W, the pullback $f^{-1}(W^\vee)$ is either empty or has codimension k, by the standard translation argument [K, Proposition 1.1]. (In characteristic 0, the same result shows that $f^{-1}(W^\vee)$ is generically smooth for general W.) We call M_k the kth *polar locus* relative to W on X, and we define it for any W.

Set $s_\alpha = $ component of dimension α of the Segre class of J in X, and write $h \in \mathrm{Pic}(X)$ for the hyperplane class. Although M_k does not come with a natural scheme structure, its construction gives a canonical cycle class on X whose support is M_k, namely $[M_k] = \pi_* f^* [W^\vee]$, which is called the kth *polar class*. The following description of $[M_k]$ is a main goal of the theory. We have

$$(1.1) \qquad [M_k] = (d - 1)^k h^k \cap [X] - \sum_{i=0}^{k-1} \binom{k}{i} (d-1)^i h^i \cap s_{n-k+i},$$

in the cycle class group $A_{n-k}(X)$; this result is [**KO**; IV, p.48] and [**P**, (2.3)]. By 1.1, the class $[M_k]$ is independent of W.

To introduce our main constructions, we now indicate a proof, which recasts the argument of [**F**, pp.8-85]. Define the *superpolar subscheme*

$$N_k = N_k(X, W) = f^{-1}(W^\vee),$$

in $\tilde{X}$, and view this as the primary object (for any W, in any characteristic). Unlike its image M_k, the superpolar has a canonical scheme structure. Set

$$\lambda = c_1(\mathscr{O}_{\mathbf{P}^{n+1}}(d - 1)) \qquad \text{and} \qquad \epsilon = c_1(\mathscr{O}_{\tilde{X}}(E)) = c_1(\mathscr{O}_{\tilde{X}}(-1)),$$

where, as usual, E denotes the exceptional divisor. Then, using the standard relation

$$f^*\mathscr{O}_{\tilde{\mathbf{P}}^{n+1}}(1) = \mathscr{O}_{\tilde{X}}(-E) \otimes \pi^*\mathscr{O}_{\mathbf{P}^{n+1}}(d - 1),$$

which holds because the linear system defining f is spanned by sections of $\mathscr{O}(d-1)$, we obtain:

$$[N_k] = (\pi^*\lambda - \epsilon)^k \cap [\tilde{X}]$$

$$= \pi^*\lambda^k \cap [\tilde{X}] - \sum_{i=0}^{k-1} \binom{k}{i} \pi^*\lambda^i \cap (-1)^{k-i-1}[E]^{k-i}.$$

Define the *total superSegre class*

$$\tilde{s}(J, X) = \sum_{i \geq 1} (-1)^{i-1}[E]^i,$$

so we have

$$s(J, X) = \pi_*\tilde{s}(J, X),$$

by [**F**, Corollary 4.2, p.75].

Rewriting the last expression for $[N_k]$, we obtain immediately a key identity:

$$(1.2) \qquad [N_k] = \pi^*\lambda^k \cap [\tilde{X}] - \sum_{i=0}^{k-1} \binom{k}{i} \pi^*\lambda^i \cap \tilde{s}_{n-k+i}(J, X).$$

Since $\pi_*[N_k] = [M_k]$, equation (1.1) follows by the projection formula.

The next step is to refine the polar classes. The subspace $W \subset \mathbf{P}^{n+1}$ is the span of k independent points $p_0, \ldots, p_{k-1}$. Denote by DF_i the directional derivative of F from p_i, set $D_i = V(DF_i)$, and let $Q \subset \mathbf{P}^{n+1}$ be the intersection of the D_i. As point sets, we have $Q \cap X = M_k \cup J$. We obtain a refined version of (1.2) by replacing suitable factors λ on the right-hand side by the classes $[D_i]$. We find

$$(1.3) \qquad [N_k] = \pi^* \prod_{i=0}^{k-1}[D_i] - \sum_{j=0}^{k-1} \binom{k}{j} \pi^*\lambda^j \cap \tilde{s}_{n-k+j}(J, X)$$

in $A_*(\pi^{-1}(Q \cap X))$.

The interpretation of (1.3) is clear: the product on the right represents the total-intersection of the divisors D_i, while the sum which follows represents the part of that intersection carried by the singularities of X. To illustrate, suppose the singularities of X are isolated. Denote by d' the class of X; as before, we shall write d for the degree. Take $k = n = \dim(X)$, and push down to X: we obtain Plücker's formula,

$$(1.4) \qquad d' = d(d-1)^n - \sum_{i=1}^{r} e_J(P_i, X),$$

from an identity in $A_0(Q \cap X)$. Since $Q \cap X$ is discrete, the latter is a *cycle* identity. For general X, one obtains the standard formula [F, p.84] for the degree of the dual variety from a cycle identity.

2. Families of hypersurfaces

Throughout this section we shall study a flat 1-parameter family of hypersurfaces in $\mathbf{P}^{n+1}$, parametrized by a smooth curve T, with total space

$$X \hookrightarrow \mathbf{P}_T^{n+1}.$$

Here we assume that the general X_t is generically reduced, hence has a conormal scheme. Again we have $X = V(F)$, at least locally on T, and this time we write J for the relative Jacobian locus $\mathrm{Jac}(X/T)$, that is, for the zero scheme of $\partial F/\partial X_0, \ldots, \partial F/\partial X_{n+1}$. We shall choose a special point $0 \in T$, and, for any $t \in T$, we shall write X_t for the fiber over t. Again we have a diagram

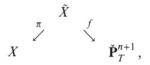

this time of T-schemes, where $\tilde{X}$ is now the closure of the graph of the Gauss map on the relative smooth locus $\mathrm{sm}(X/T)$, that is, the conormal scheme of the family $\{X_t\}$, so that f_t is the Gauss map of X_t, when t is general.

For the next result, we shall assume that every X_t has isolated singularities. In this case, X_t has a conormal scheme, and the polars and superpolars of X_t are well-defined, for every $t \in T$. We shall write CX_t (resp. J_t) for the conormal scheme (resp. the Jacobian subscheme) of X_t, and CX for $\tilde{X}$. Further, we denote by Q_t the subscheme Q as above, defined for each X_t; these, too, travel in an algebraic family.

Under these conditions, observe that the cycle of $\lim_{t \to 0} CX_t$ takes the simple form

$$[(CX)_0] = [C(X_0)] + \sum_i m_i[CP_i],$$

where the m_i are nonnegative integers and $\{P_1, \ldots, P_k\} = \text{Supp}(J(X_0))$. Indeed, by definition of the conormal scheme, the term $[C(X_0)]$ appears, and any other components must lie over $\text{Supp}(J)$. But, over each P_i, we must have either nothing new, or else a component of dimension $n = \dim(X_0)$. Because $(CX)_0 \subset I$, the only possibility is therefore a component of the form $m_i[CP_i]$, as asserted.

To determine the m_i, we take superpolars. Choose a linear subspace $W = W_n$ of dimension $n-1$ in $\mathbf{P}^{n+1}$. From the last relation we obtain

$$(2.1) \qquad \lim_{t \to 0}[N_n(X_t, W)] = [N_n(X_0, W)] + \sum_i m_i[CP_i] \cap [W^\vee].$$

Here $W^\vee$ is a line, so the intersection $[CP_i] \cap W^\vee$ is a reduced point $(P_i, H_i) \in I$.

It will be convenient, for a zero-cycle $Z = \sum \alpha_i P_i$ in X, to write $CZ = \sum \alpha_i CP_i$ for the pullback cycle on the incidence correspondence.

THEOREM 2.2 (Van Gastel). *For a family of hypersurfaces as above, with isolated singularities, we have*

$$[(CX)_0] = [C(X_0)] + \left[C\left(s_0(J_0, X_0) - \lim_{t \to 0} s_0(J_t, X_t)\right)\right],$$

as n-cycles on CX.

PROOF. All we need to do is to show that the m_i are as indicated by the the second term on the right. For any $t \in T$, we have

$$\pi_*\left([N_n(X_t, W)] + \tilde{s}_0(J_t, X_t)\right) = Q_t.X_t,$$

hence

$$\pi_* \lim_{t \to 0}\left([N_n(X_t, W)] + \tilde{s}_0(J_t, X_t)\right) = \lim_{t \to 0} Q_t.X_t$$

in $A_0(Q_0 \cap X_0)$. Indeed, π_* and $\lim_{t \to 0}$ commute, by [**F**, Proposition 10.1(a), p.177], which asserts that specialization of cycles commutes with proper pushforward, and [**F**, Proposition 11.1(b), p.196], which interprets limits as specializations. But

$$\pi_*\left([N_n(X_0, W)] + \tilde{s}_0(J_0, X_0)\right) = Q_0.X_0,$$

also in $A_0(Q_0 \cap X_0)$. Because X_0 has isolated singularities, so that $Q_0 \cap X_0$ is discrete, it follows that the identities above hold at the cycle level. We also have

$$\lim_{t \to 0} Q_t.X_t = Q_0.X_0,$$

as zero-cycles on X, by continuity. Hence we obtain the cycle identity

$$(2.3) \qquad \lim_{t \to 0}[N_n(X_t, W)] = [N_n(X_0, W)] + \tilde{s}_0(J_0, X_0) - \lim_{t \to 0} \tilde{s}_0(J_t, X_t) + \epsilon,$$

where $\pi_* \epsilon = 0$ in $A_0(Q_0 \cap X_0)$. Now compare (2.3) and (2.1): the theorem follows.

Examples. Suppose a nonsingular plane cubic X degenerates to a reduced limit curve X_0. If X_0 is nodal, the conormal scheme CX_0 is of the form $[CX_0] + 2[CP]$ where P is the node. In this case we say that the limit X_0 has a *double vertex* at P. The limit dual curve, a sextic, contains the line dual to P with multiplicity 2. If X_0 is cuspidal, the limit conormal scheme is $[CX_0] + 3[CP]$, where P is the cusp. Here X_0 has a *triple vertex* at P, and the limit dual curve contains the line dual to P with multiplicity 3. In both cases, the limit term to be subtracted on the right in Van Gastel's formula is zero.

For an example with a nontrivial subtraction, suppose a nodal cubic X degenerates to a cuspidal X_0. Then the limit conormal scheme is $[CX_0] + [CP]$, where P is the cusp. Here X_0 has a *simple vertex* at P, and the limit dual curve, a quartic, contains the line dual to P with multiplicity 1.

When the limit is reduced, the multiplicity of a vertex reflects the local change in the Jacobian multiplicity as $X \to X_0$. When the limit is nonreduced, however, the interplay with the parametrization is more complex.

3. Nonreduced special fibers

Now consider a flat 1-parameter family of curves as in the last section, but in $\mathbf{P}^2$, where the general X_t is reduced but the special fiber X_0 is nonreduced. Denote by $G \in \Gamma(X_0, N_{X_0}\mathbf{P}^2)$ a characteristic section of the family [F, 11.3], by $Z(G)$ its subscheme of zeros, and by R the subscheme residual to $X_0 - (X_0)_{\mathrm{red}}$ in the Jacobian subscheme J_0 of X_0.

THEOREM 3.1 (Van Gastel). *In the situation above, assume that $Z(G)$ has dimension 0. Then we have*

$$[(CX)_0] = [C(X_0)] + C[Z(G).([X_0] - [(X_0)_{\mathrm{red}}]) + s(R, X_0) - \lim_{t \to 0} s(J_t, X_t)].$$

The proof is based on Lazarsfeld's analysis ([L, F, Example 11.3.2]) of the excess intersection of two plane curves. It generalizes Sheldon Katz' earlier discussion [Ka] of the degenerate members of a pencil.

Example 1. Suppose a pencil of elliptic curves degenerates to a triple line. Then, by 3.1, the conormal scheme degenerates to a triple line with 3 double vertices.

Example 2. Let X be a nonsingular cubic in $\mathbf{P}^2$, and choose a point $P \in \mathbf{P}^2$ such that the 6 tangents from P to X are distinct. Fix a line L off P, and denote by $V_1, \ldots, V_6$ the intersections of the 6 tangents with L. Finally, choose coordinates so that L is the x-axis and P is at ∞ on the y-axis. The family of linear maps defined in affine coordinates by $(x, y) \to (x, ty)$, parametrized by $t \in A^1$, defines a family of cubics X_t, such that $X_1 = X$ and X_0 is the tripled line L. The conormal scheme cycle $[C(X_t)]$ degenerates to

$$3[L] + [CV_1] + \cdots + [CV_6].$$

Indeed, denote by π the projection from P. The 6 given tangents meet X_t in the polar locus $M_1(X_t, P)$. For $t \neq 0$, the y-discriminant cuts out the image $\pi(M_1(X_t, P))$ on L, which is $\{V_1, \ldots, V_6\}$. Passing to the limit, we see that the lines from the V_i to P are in $(CX)_0$. Varying P, it follows that $V_1, \ldots, V_6$ are vertices on the triple line. (For a generalization, see [**VG**, Proposition 5.1].) Because the dual curve is a sextic, these vertices are simple. By 3.1, the characteristic section must be identically 0. In fact from [**KS**], one can see directly that the degeneration is second-order.

The case of a triple line with distinct vertices is crucial for the enumeration of nonsingular plane cubics. In fact, limits of this kind appear in the so-called *elementary systems*. These are defined as follows. Let α and β be nonnegative integers such that $\alpha + \beta = 8$. Choose α general points in $\mathbf{P}^2$, and β general lines. Denote by $T = T_{\alpha, \beta}$ the normalization of the closure of the curve in $\mathbf{P}^9 = \{$ plane cubics $\}$ parametrizing the smooth cubics through the α points, tangent to the β lines. The conormal family $\{X_t\}_{t \in T}$ is called the *elementary system* given by α points and β lines. For $\alpha \leq 2$, the elementary system $\{X_t\}$ contains finitely many triple lines, each with a limit conormal schemes equipped with 6 *distinct* vertices. The characteristic numbers for nonsingular plane cubics can be found efficiently by passing first to the elementary systems, and then counting, more or less directly, the degenerate curves which they contain. (For details, see [**KS**].)

This example shows that Theorem 3.1 is not general enough for some important applications. In [**KS**], to locate the vertices on nonreduced degenerate cubics, we avoided this shortcoming by taking discriminants, generalizing the procedure of Example 2. While the resulting calculations confirm strikingly the emphasis placed by Schubert and Zeuthen on the degenerations which arise from homolographies, it would nonetheless be far better to obtain such results from a stronger dynamic intersection theory.

4. Local complete intersections: preliminaries

To generalize the situation in §1, suppose that our given variety X is a generically smooth local complete intersection of dimension n in $\mathbf{P}^{n+r}$, defined by the sheaf $\mathscr{F}$ of $\mathscr{O}_{\mathbf{P}^{n+r}}$-ideals. The conormal sheaf $\mathscr{F}/\mathscr{F}^2$ is locally free of rank r, and we have the standard presentation

$$\mathscr{F}/\mathscr{F}^2 \xrightarrow{\alpha} \Omega^1_{\mathbf{P}^{n+r}}|_X \longrightarrow \Omega^1_X \to 0.$$

For a point $x \in X$, the reduction α_x is injective exactly when the fiber $\Omega^1_X \otimes k(x)$ has rank n, in other words when X is smooth at x. Denote by J the singular locus of X. (We could give J a scheme structure via the Fitting ideal $F^n \Omega^1_X$, but in §6 we shall define the scheme structure in another way, more natural for our purposes. For now, we only need to know that J is closed.) For a function $F \in \mathscr{F}$ vanishing to first order on X, the class of F maps under α to the restriction $dF|_X$ of its differential.

To dominate the conormal scheme of X, take $\mathbf{P}^{n+r} = P(V)$, for a vector space V of dimension $n + r + 1$ over the base field k. Then the standard exact sequence for the 1-forms on $\mathbf{P}^{n+r}$ reads

$$0 \to \Omega^1_{\mathbf{P}^{n+r}} \xrightarrow{\beta} \mathscr{O}_{\mathbf{P}^{n+r}}(-1) \otimes_k V^\vee \longrightarrow \mathscr{O}_{\mathbf{P}^{n+r}} \to 0,$$

where $V^\vee$ denotes the dual vector space. In particular, the map β is the inclusion of a subbundle. Composing α with the inclusion $\beta|_X$, we obtain a map of locally free sheaves

$$\mathscr{F}/\mathscr{F}^2 \xrightarrow{\gamma} \mathscr{O}_X(-1) \otimes_k V^\vee,$$

on X, which is injective (onto a subbundle) exactly on X_{reg}. If we choose coordinates $X_0, \ldots, X_{n+r}$ in $\mathbf{P}^{n+r}$, the class of a function $F \in \mathscr{F}$ vanishing to first order on X is mapped by γ to the restriction $\nabla F|_X = (\partial F/\partial X_0, \ldots, \partial F/\partial X_{n+r})|_X$ of its gradient.

Write C for the projective bundle associated to the conormal sheaf $\mathscr{F}/\mathscr{F}^2$, with structural projection

$$C = P(\mathscr{F}/\mathscr{F}^2) \xrightarrow{p} X.$$

A smooth point of C may be viewed as a point $x \in X_{\text{reg}}$, equipped with a hyperplane $H \subset \mathbf{P}^{n+r}$ containing the embedded tangent space $T_x X$, where H is represented by the germ at x of a function $F \in I$, unique modulo I^2. It follows that $p^{-1}(X_{\text{reg}})$ is the conormal scheme of X_{reg}. From γ we obtain a rational map

$$P(\mathscr{F}/\mathscr{F}^2) \xrightarrow{\rho} P(\mathscr{O}_X(-1) \otimes_k V^\vee),$$

whose target is canonically isomorphic to $X \times \check{\mathbf{P}}^{n+r}$, where the second factor is the dual projective space. In fact, for a smooth point x, we have $\rho(F) = (x, H)$. Hence, by definition [K], the conormal scheme CX is the closure of the image of $\rho|X_{\text{reg}}$ in $X \times \check{\mathbf{P}}^{n+r}$.

Denote by

$$C \xrightarrow{\varphi} \check{\mathbf{P}}^{n+r},$$

the rational map obtained by composing ρ with the second projection of $X \times \check{\mathbf{P}}^{n+r}$, and write B for the base locus of the linear system which defines φ. A first step toward understanding its geometry is

PROPOSITION 4.1. *The projection $p : C \to X$ maps B onto J. (In particular, when $r = 1$, we have $B = J$ under the natural identification of C and X.)*

PROOF. The question is local, so we can pass to an open affine U on X. On U we have $B = \{(x, F)| (\nabla F)_x = 0\}$, for $x \in X$ and $F \in \mathscr{F}/\mathscr{F}^2$. Localizing again if necessary, we can assume that $\mathscr{F}/\mathscr{F}^2$ is spanned by

$F_1, \ldots, F_r$. Denote by M_x the Jacobian matrix $(\partial F_i / \partial x_j)|_x$, where $i = 1, \ldots, r$ and $j = 0, \ldots, n + r$. Then

$$J = \{x \mid \operatorname{rank}(M_x) < r\} = \{x \mid (\nabla F)_x = 0 \text{ for some } F \in \mathscr{I}/\mathscr{I}^2\},$$

and the proposition follows immediately. (For another proof, see the remark after 6.1 below.)

Now we introduce our fundamental construction. Denote by $\tilde{X}$ the blow-up of C along B, with structural map $\tilde{X} \xrightarrow{q} C$. Then φ induces a morphism $f: \tilde{X} \to \check{\mathbf{P}}^{n+r}$, and we have a diagram

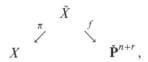

where π is the composite $p \circ q$. By the definition of the conormal scheme CX, the projection

$$C \times \check{\mathbf{P}}^{n+r} \to X \times \check{\mathbf{P}}^{n+r},$$

maps $\tilde{X}$ onto CX. The induced projection

$$\tilde{X} \xrightarrow{\psi} CX,$$

is birational, in fact biregular over X_{reg}. (In the special case $r = 1$, it is easy to see that this diagram reduces to its counterpart in §1.)

Set

$$\mathscr{L} = \mathscr{O}_C(1) \otimes p^* \mathscr{O}_X(-1) \in \operatorname{Pic}(C),$$

and denote by E the exceptional divisor of the blowup $\tilde{X}$.

PROPOSITION 4.2. *In* $\operatorname{Pic}(\tilde{X})$, *we have*

$$f^* \mathscr{O}_{\check{\mathbf{P}}^{n+r}}(1) = q^*(\mathscr{L}) \otimes \mathscr{O}_{\tilde{X}}(-E) = q^* \mathscr{L} \otimes \mathscr{O}_{\tilde{X}}(1).$$

In particular, if $r = 1$, and $X = V(F)$ is a hypersurface of degree d, then p is an isomorphism, and the standard identification [F, B.5.5, p.434] of X and C gives $\mathscr{O}_C(1) = p^* \mathscr{O}_X(d)$, so that $\mathscr{L} = p^* \mathscr{O}_X(d-1)$, as in §1.

PROOF. First we verify 4.2 when X is a smooth locally closed subscheme of $\mathbf{P}^{n+r}$, so that the rational map p above is an embedding, and $\pi = p$.

Write $F = \mathscr{O}_X(-1) \otimes_k V^{\vee}$, and $G = \mathscr{O}_X \otimes_k V^{\vee}$, the latter a trivial bundle. Then f factors into the composite

$$C \xrightarrow{p} P(F) \xrightarrow{\sim} P(G) \to \check{\mathbf{P}}^{n+r},$$

where the middle arrow is the standard isomorphism [F, B.5.5, p.434]. The tautological line bundle $\mathscr{O}_{\check{\mathbf{P}}^{n+r}}(1)$ pulls back to $\mathscr{O}_G(1)$ on the trivial bundle $P(G)$, and then, by [F, loc. cit.], to $\mathscr{O}_F(1) \otimes \phi^* \mathscr{O}_X(-1)$, where $\phi: P(F) \to X$ is the structure map. But $\sigma^* \mathscr{O}_F(1) = \mathscr{O}_C(1)$, because $\mathscr{I}/\mathscr{I}^2$ is a subbundle of F, and $\sigma^* \phi^* = p^*$ so that $\phi^* \mathscr{O}_X(-1)$ pulls back to $p^* \mathscr{O}_X(-1)$. Hence we find

$$f^* \mathscr{O}_{\check{\mathbf{P}}^{n+r}}(1) = \mathscr{O}_C(1) \otimes p^* \mathscr{O}_X(-1) = \mathscr{L},$$

which verifies the proposition when X is smooth.

For a singular X, the preceding argument describes the pullback of $\mathscr{O}_{\check{\mathbf{P}}^{n+r}}(1)$ to $C - B$. Because the sections of $f^* \mathscr{O}_{\check{\mathbf{P}}^{n+r}}(1)|_{p^{-1}(X_{\text{reg}})}$ extend by 0 over B to give sections of $\mathscr{L}$, the well-known expression for $f^* \mathscr{O}_{\check{\mathbf{P}}^{n+r}}(1)$ on the blowup $\tilde{X}$ completes the proof.

5. Local complete intersections: superpolars

Let k be an integer, with $0 \leq k \leq n$, and suppose given a linear subspace W_k of dimension $k + r - 2$ in $\mathbf{P}^{n+r}$. We denote by $W^\vee$ the dual subspace:

$$W^\vee = \left\{ H \in \check{\mathbf{P}}^{n+r} \mid H \supset W_k \right\}.$$

Set

$$M_k = \text{closure in } X \text{ of } \left\{ p \in X_{\text{reg}} \mid T_p X \text{ and } W_k \text{ span at most a hyperplane} \right\}$$

$$= \text{closure in } X \text{ of } \left\{ p \in X_{\text{reg}} \mid \dim(T_p X \cap W_k) > k - 2 \right\}.$$

As a point set, we have $M_k = \pi f^{-1}(W^\vee)$, where $\tilde{X} \xrightarrow{f} \check{\mathbf{P}}^{n+r}$ is the natural projection. We call M_k the kth *polar locus* of X relative to W_k. Define the *superpolar subscheme*:

$$N_k = N_k(X, W) = f^{-1}(W^\vee),$$

in $\tilde{X}$. In particular, N_k maps onto M_k. For general W, the subscheme N_k has codimension exactly $k + r - 1$ in $\tilde{X}$, by the standard translation argument. (In characteristic 0, it also follows that N_k is generically smooth for general W.) Our plan is to describe $[N_k]$ first, using appropriate superSegre classes, and then define $[M_k]$ as a direct image.

PROPOSITION 5.1. *For general* W_k, *the superpolar subscheme* $N_k \in \tilde{X}$ *projects to its image in* X *by a generic isomorphism. In particular,* M_k *has codimension* k *in* X.

PROOF. Since W_k is general, the standard translation argument shows that each N_k meets the exceptional divisor of $\tilde{X}$ properly, and therefore contains a dense open set of points away from the exceptional divisor. Passing to this open set, we may assume that X is a smooth locally closed subscheme of $\mathbf{P}^{n+r}$, and hence that $\tilde{X} = C$. We begin with N_0. For a general linear subspace $W \subset \mathbf{P}^{n+r}$ of dimension $r - 2$, and for a general $x \in X$, it is easy to see that W and the embedded tangent space $T_x X$ span a unique hyperplane, so that the projection $N_0 \to X$ is generically bijective, at least. To go further, set $t = c_1(\mathcal{O}_C(1))$ and set $\lambda = c_1(\mathcal{L})$ in $\text{Pic}(C)$, and $h = c_1(\mathcal{O}_X(1))$ in $\text{Pic}(X)$. By definition of $\mathcal{L}$, we have $\lambda = t - p^*h$, hence

$$
\begin{aligned}
p_*[N_0] &= p_*((t - p^*h)^{r-1} \cap [C]) \\
&= p_* \left((t^{r-1} - (r-1)t^{r-2}p^*h + \cdots) \cap [C] \right) \\
&= p_*(t^{r-1} \cap [C]) \\
&= s_n(X, \mathbf{P}^{n+r}),
\end{aligned}
$$

using the projection formula, commutativity of $\cap$, the definition of $s_n(X, \mathbf{P}^{n+r})$, and the obvious identity

$$
p_*(t^i \cap p^*(\alpha)) = 0,
$$

for $i < r-1 = \dim(C/X)$, all $\alpha \in A(X)$. But the generically smooth X has multiplicity $e_X \mathbf{P}^{n+r} = 1$ in $\mathbf{P}^{n+r}$, so [**F**, 4.3, p.79] gives $p_*[N_0] = [X]$. This shows that the projection $p|N_0 : N_0 \to X$ is birational. We obtain $[N_k]$ from $[N_0]$ by intersecting with further pullbacks of general hyperplanes; these meet properly, and their intersection meets properly the closed set where $p|N_0$ is not an isomorphism, by the usual translation argument. Hence each $[N_k]$ projects to X by a generic isomorphism, as asserted.

Using 4.2, we can now calculate the superpolar classes. As in §1, we shall study them first on $\tilde{X}$, and then refine them. Denote by E the exceptional divisor on $\tilde{X}$, and set

$$
\lambda = c_1(\mathcal{L}) \qquad \epsilon = c_1(\mathcal{O}_{\tilde{X}}(E)) = c_1(\mathcal{O}_{\tilde{X}}(-1)).
$$

Because $W^\vee$ is the intersection of $k + r - 1$ hyperplanes in $\check{\mathbf{P}}^{n+r}$, using 4.2 we obtain

$$
\begin{aligned}
[N_k] &= (q^*\lambda - \epsilon)^{k+r-1} \cap [\tilde{X}] \\
&= q^*\lambda^{k+r-1} \cap [\tilde{X}] - \sum_{i=0}^{k+r-2} \binom{k+r-1}{i} q^*\lambda^i \cap (-1)^{k+r-i}[E]^{k+r-1-i},
\end{aligned}
$$

in $A_{n-k}(\tilde{X})$.

We define the *total superSegre class* exactly as we did in §1:

$$
\tilde{s}(B, C) = \sum_{i \geq 1} (-1)^{i-1}[E]^i.
$$

Again, we have

$$
q_*\tilde{s}(B, C) = s(B, C),
$$

by [**F**, Corollary 4.2.2, p.75]. From the last expression for $[N_k]$, we obtain

$$
(5.2) \quad [N_k] = q^*\lambda^{k+r-1} \cap [\tilde{X}] - \sum_{i=0}^{k+r-2} \binom{k+r-1}{i} q^*\lambda^i \cap \tilde{s}_{n-k+i}(B, C),
$$

in $A_{n-k}(\tilde{X})$.

Next, we refine the superpolar classes. The subspace W_k of $\mathbf{P}^{n+r}$ is the span of $k + r - 1$ independent points $p_0, \ldots, p_{k+r-2}$. For each $i = 0, \ldots, k - r - 2$, denote by $\check{H}_i$ the hyperplane in $\check{\mathbf{P}}^{n+r}$ dual to p_i. As we observed in the proof of 4.2, sections of $\mathscr{O}_{\check{\mathbf{P}}^{n+r}}(1)$, in particular those defining the H_i, pull back to C, yielding divisors D_i on C. Denote by Q the intersection scheme of the D_i. Clearly the support of Q lies in $q(N_k) \cup B$, so we obtain a refined version of 5.2 by replacing the factors λ in the first term on the right-hand side by the classes $[D_i]$. We obtain

THEOREM 5.3. *Notations as above, we have*

$$[N_k] = q^* \prod_{i=0}^{k+r-1} [D_i] - \sum_{i=0}^{k+r-2} \binom{k + r - 1}{i} q^* \lambda^i \cap \tilde{s}_{n-k+i}(B, C),$$

in $A_{n-k}(q^{-1}Q)$.

Taking direct images on C, we obtain the following result, which reduces to 1.1 when $r = 1$.

COROLLARY 5.4. *Notations as above, we have*

$$q_*[N_k] = \prod_{i=0}^{k+r-1} [D_i] - \sum_{i=0}^{k+r-2} \binom{k + r - 1}{i} \lambda^i \cap s_{n-k+i}(B, C),$$

in $A_{n-k}(Q)$.

6. Local complete intersections: polars

Continue with the previous notation and assumptions, and denote by δ the dual of the natural map

$$\mathscr{I}/\mathscr{I}^2 \xrightarrow{\alpha} \Omega^1_{\mathbf{P}^{n+r}}|_X.$$

Then, for Lichtenbaum and Schlessinger's cotangent functor T^1, we have

$$T^1(X, \mathscr{O}_X) = \operatorname{cok} \delta,$$

by [LS, 2.2.4 and 3.1.1]. Write $\mathbf{T}^1(X)$ for the scheme $\operatorname{Proj}(\operatorname{Symm}(T^1(X, \mathscr{O}_X))$. It follows from [LS, 2.1.9] that $\mathbf{T}^1(X)$ depends only on X, and not on its projective embedding.

PROPOSITION 6.1. *We have a canonical isomorphism*

$$B \cong \mathbf{T}^1(X),$$

of C-*schemes.*

PROOF. Recall that the rational map φ of §4 is induced by δ, and that B is the corresponding base locus in $C = P(\mathscr{I}/\mathscr{I}^2)$. In other words, we have $B = \operatorname{Proj}(\operatorname{Symm}(\operatorname{cok}\delta))$, so the proposition follows immediately.

Since the support of $T^1(X, \mathscr{O}_X)$ is the singular locus J, we use the $\mathscr{O}_X$-ideal sheaf $\text{Ann}(T^1(X, \mathscr{O}_X))$ to define a scheme structure on J. In the special case $r = 1$, when X is a hypersurface, the scheme structure just defined for J agrees with that of §1. We also recover 4.1 as a corollary of the proposition.

For a linear subspace W_k of dimension $k + r - 2$ of $\mathbf{P}^{n+r}$, we define the kth *polar class* to be

$$[M_k] = [M_k(X, W)] = \pi_*[N_k(X, W)],$$

in $A_{n-k}(M_k \cup J)$, where $\tilde{X} \overset{\pi}{\to} X$ is the projection.

To compute $[M_k]$, push forward via the inclusion $M_k \cup J \hookrightarrow X$. By 5.4, writing classes now on X, we have

$$[M_k] = \mathfrak{a} - \mathfrak{b},$$

where

$$\mathfrak{a} = p_* \prod_{i=0}^{k+r-1} [D_i] \quad \text{and} \quad \mathfrak{b} = p_* \sum_{i=0}^{k+r-2} \binom{k+r-1}{i} \lambda^i \cap s_{n-k+i}(B, C).$$

To find $\mathfrak{a}$, we have

$$\mathfrak{a} = p_*(\lambda^{k+r-1} \cap [C])$$
$$= p_*(\lambda^{k+r-1} \cap p^*[X])$$
$$= p_*((t - p^*h)^{k+r-1} \cap p^*[X])$$
$$(6.2) \qquad = \sum_{i=0}^{k+r-1} (-1)^i \binom{k+r-1}{i} p_*(t^{k+r-1-i} \cap p^*(h^i \cap [X]))$$
$$= \sum_{i=0}^{k+r-1} (-1)^i \binom{k+r-1}{i} s_{k-i}(\mathscr{F}/\mathscr{F}^2) \cap h^i \cap [X].$$

As in §5, we have written $\lambda = t - p^*h$, with $t = c_1(\mathscr{O}_C(1))$ and $h = c_1(\mathscr{O}_X(1))$, by definition of $\mathscr{L}$. Denote by s_{k-i} the Segre class operator of [F, Chapter 3].

Now assume that X has isolated singularities, at the closed points $p_1, \dots, p_s$. To work out $\mathfrak{b}$, denote by J_i the connected component of J supported by p_i, and choose an open affine neighborhood U_i of J_i on which the projective bundle C is trivial. Set $B_i = B|_{J_i}$, and write C_i for the restriction of C to U_i. One checks directly that the isomorphism $C_i \cong U_i \times \mathbf{P}^{r-1}$ identifies B_i with $J_i \times L_i$, for a suitable linear subspace L_i in $\mathbf{P}^{r-1}$. By [F, Example 4.2.5, p.77], we obtain

$$(6.3) \qquad s(B, C) = \sum_{i=1}^{s} s(B_i, C) = s(J_i, X) \times s(L_i, \mathbf{P}^{r-1}).$$

Under our assumption that X has isolated singularities, the only relevant $\mathfrak{b}$ is when $k = n$. To compute it, 6.3 gives

$$\lambda^i \cap s_i(B_j, C_j) = s_0(J_j, X) \times (t^i \cap s_i(L_j, \mathbf{P}^{r-1})).$$

Indeed, for $\alpha \in A_*(J_j)$ and $\beta \in A_*(L)$, we have $t \cap (\alpha \times \beta) = \alpha \times (t \cap \beta)$, and $p^* h \cap (\alpha \times \beta) = (h \cap \alpha) \times \beta = 0$. Now set

$$d_{i,j} = \int_{\mathbf{P}^{r-1}} t^i \cap s_i(L_j, \mathbf{P}^{r-1})$$

$$= \text{coefficient of } t^{\dim(L_j)} \text{ in } t^i/(1+t)^{r-1-\dim(L_j)},$$

and set

$$d_j = \sum_{i=0}^{n+r-2} \binom{n+r-1}{i} d_{i,j}.$$

Then, by the definition of $\mathfrak{b}$, we obtain

$$(6.4) \qquad \mathfrak{b} = \sum_{j=1}^{s} d_j s_0(J_j, X) = \sum_{j=1}^{s} d_j (e_{J_j} X)[p_j].$$

In combination, 6.2 and 6.4 give:

THEOREM 6.5. *Assume that X has isolated singularities. Then we have*

$$[M_k] = \mathfrak{a} - \mathfrak{b},$$

in $A_{n-k}(M_k \cup J)$, where

$$\mathfrak{a} = \sum_{i=0}^{k+r-1} (-1)^i \binom{k+r-1}{i} s_{k-i}(\mathscr{I}/\mathscr{I}^2) \cap h^i \cap [X],$$

and

$$\mathfrak{b} = \begin{cases} \sum_{j=1}^{s} d_j (e_{J_j} X)[p_j] & \text{if } k = n; \\ 0 & \text{otherwise.} \end{cases}$$

7. Limit local complete intersections

To generalize Van Gastel's theorem 2.2, we shall suppose given a flat 1-parameter family of local complete intersections of dimension n in $\mathbf{P}^{n+r}$, parametrized by a smooth curve T, with total space

$$X \hookrightarrow \mathbf{P}_T^{n+r}.$$

Here we assume, for general $t \in T$, that the corresponding fiber X_t is generically reduced, hence has a conormal scheme.

Denote by $\mathscr{I}$ the ideal sheaf of X in $\mathscr{O}\mathbf{P}_T^{n+r}$, so that the conormal sheaf $\mathscr{I}/\mathscr{I}^2$ is locally free of rank r. Again write C for the associated projective bundle, with structural projection $C = P(\mathscr{I}/\mathscr{I}^2) \xrightarrow{p} X$. We obtain as before a rational map

$$C \xrightarrow{\varphi} \check{\mathbf{P}}_T^{n+r},$$

whose fiber over a closed point of T identifies with the previous φ defined for a single variety in §4.

Denote by B the base locus of the linear system (relative to T) which defines φ, and denote by J the relative singular locus of X/T, as a point set. The constructions commute with base change, hence 4.1 implies that B maps onto J. Denote by $\tilde{X}$ the blowup of C along B, with structural map

$$\tilde{X} \xrightarrow{q} C.$$

As before, we have a diagram

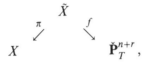

this time of T-schemes, which identifies $\tilde{X}$ with the closure of the graph of the rational map φ. Further, $\tilde{X}$ dominates the conormal scheme, denoted CX, of the family $\{X_t\}$, by a birational morphism $\tilde{X} \xrightarrow{\psi} CX$, which is biregular on the pullback of the relative smooth locus $\operatorname{sm}(X/T)$.

For the next result, we shall assume again that every X_t has isolated singularities. In this case, each X_t has a conormal scheme, and the polars and superpolars of X_t are well defined, for every $t \in T$. Clearly, for general t, we have $(CX)_t = C(X_t)$.

Now fix a closed point $0 \in T$, and, for any $t \in T$, write X_t for the fiber over t. Then the fiber f_t is the map of $X_t \to \check{\mathbf{P}}^{n+r}$ of §4. Our goal, as in §2, will be to compare the fiber

$$(CX)_0 = \lim_{t \to 0} C(X_t),$$

with the conormal scheme $C(X_0)$.

Because of our assumption on the singularities, the cycle of $(CX)_0$ takes the simple form

(7.1) $$[(CX)_0] = [C(X_0)] + \sum_i m_i [CP_i],$$

where the m_i are nonnegative integers and $\{P_1, \dots, P_k\} = \operatorname{Supp}(J(X_0))$, by the same reasoning (see §2) as for a family of hypersurfaces.

To state our main result, suppose that the Jacobian subscheme of the general X_t is supported by the closed points $P_{1,t}, \dots, P_{r,t} \in X$. For $i = 1, \dots, r$, denote by J_i the component of J supported at P_i, and set

$$\alpha_{i,t} = d_i e_{J_i} X_t,$$

with d_i as in 6.4, where $e_{J_i} X_t$ is the multiplicity of J_i in X_t, that is, the multiplicity, in the local ring of X_t at the given point, of the Jacobian ideal.

We define the *singularity cycle* of X_t to be

$$Z_t = \sum_{i=0}^{r} \alpha_{i,t} P_{i,t}.$$

Note that the coefficients are independent of t for general t. Indeed, on the one hand, the dimension of T^1 is semicontinuous, so the d_i are constant for general t; on the other hand, the Jacobian multiplicities $e_{J_i} X$ are also semicontinuous for general t. In addition, the variation is such that each coefficient of the singularity cycle Z_0 of the special X_0 is at least as large as the corresponding coefficient of the general Z_t. The next result generalizes [**VG**, Proposition 4.5], which treats the special case of a family with smooth generic fiber.

THEOREM 7.2. *We have*

$$[(CX)_0] = [C(X_0)] + [C(Z_0 - \lim_{t \to 0} Z_t)],$$

as n-cycles on CX.

PROOF. The reasoning will be similar to that for Theorem 2.2. Because $\tilde{X}$ identifies canonically with the conormal scheme CX over the relative smooth set, we can identify the superpolars $N_k(X_t, W)$ with their direct images on CX, and similarly their cycles. Pull back a general linear subspace $W^\vee$ of codimension n from $\check{\mathbf{P}}^{n+r}$, and intersect it with the terms of (7.1). We obtain the relation

(7.3) $$\lim_{t \to 0} [N_n(X_t, W)] - [N_n(X_0, W)] = \sum_i m_i [CP_i] \cap [W^\vee].$$

For each $t \in T$, denote by Q_t the subscheme Q of §5, defined for the corresponding fiber X_t. Similarly we define B_t and C_t. By 5.3, we obtain

$$q_* \left([N_n(X_t, W)] + \sum_{i=0}^{n+r-2} \binom{n+r-1}{i} q^* \lambda^i \cap \tilde{s}_i(B_t, C_t) \right) = Q_t . X_t,$$

hence

$$q_* \lim_{t \to 0} \left([N_n(X_t, W)] + \sum_{i=0}^{n+r-2} \binom{n+r-1}{i} q^* \lambda^i \cap \tilde{s}_i(B_t, C_t) \right) = \lim_{t \to 0} Q_t . X_t,$$

in $A_0(Q_0 \cap X_0)$. Again, as in the proof of 2.2, we know that q_* and $\lim_{t \to 0}$ commute, by [**F**, Proposition 10.1(a), p.177] and [**F**, Proposition 11.1(b)], which interprets limits as specializations. But

$$q_* \left([N_n(X_0, W)] + \sum_{i=0}^{n+r-2} \binom{n+r-1}{i} q^* \lambda^i \cap \tilde{s}_i(B_0, C_0) \right) = Q_0 . X_0,$$

also in $A_0(Q_0 \cap X_0)$. We also have

$$\lim_{t \to 0} Q_t . X_t = Q_0 . X_0,$$

in $A_0(Q_0 \cap X_0)$, by continuity.

Now, because X_0 has isolated singularities, $Q_0 \cap X_0$ pushes down to a discrete subset of X, so the identities above push down to cycle identities on X. Therefore we must have a cycle identity

(7.4)
$$\lim_{t \to 0}[N_n(X_t, W)] - [N_n(X_0, W)]$$

$$= \sum_{i=0}^{n+r-2} \binom{n+r-1}{i} q^* \lambda^i \cap \left(s_i(B_0, C_0) - \lim_{t \to 0} \tilde{s}_i(B_t, C_t) \right) + \epsilon,$$

where $\pi_* \epsilon = p_* q_* \epsilon = 0$ in $A_0\left(p(Q_0 \cap X_0)\right)$. Now the sum on the right side pushes down to $Z_0 - \lim_{t \to 0} Z_t$, so, using the calculations in the last § to evaluate each term in the difference, the theorem follows by comparing (7.3) and (7.4) to determine the m_i.

References

[F] W. Fulton, *Intersection theory*, Springer-Verlag, New York, 1984.

[HMS] J. Henry and M. Merle, and C. Sabbah, *Sur la condition de Thom stricte pour un morphisme analytique complexe*, Ann. Sci. École Norm. Sup. **17** (1984), 227–268.

[Ka] S. Katz, *Tangents to a multiple plane curve*, Pacific J. Math. **124** (1986), 321–332.

[K] S. L. Kleiman, *Transversality of a general translate*, Compositio. Math. **28** (1974), 287–297.

[KO] ____, *The enumerative theory of singularities*, Real and complex singularities, Oslo 1976 (P. Holm, ed.), Sijthoff and Noordhoff 1977, 297–396.

[KC] ____, *About the conormal scheme*, Proc. Complete Intersections, Acireale 1983, Lecture Notes in Math. vol. 1092, Springer-Verlag, Berlin and New York, pp. 161–197 (1984).

[KS] S. L. Kleiman and R. Speiser, *Enumerative Geometry of nonsingular plane cubics*, these proceedings.

[L] R. Lazarsfeld, *Excess intersection of divisors*, Compositio. Math. **43** (1981), 281–296.

[LS] S. Lichtenbaum and M. Schlessinger, *The cotangent complex of a morphism*, Trans. Amer. Math. Soc., **128** (1967), 128–170.

[LT] D. T. Lê and B. Teissier, *Variétés polaires locales et Classes de Chern des variétés singulières*, Ann. of Math. **114** (1981), 457–491.

[P] R. Piene, *Polar classes of singular varieties*, Ann. Sci. École Norm. Sup. **11** (1978), 247–276.

[T] B. Teissier, *Variétés polaires*, II, Proc. La Rábida 1981, Lecture Notes in Math. vol. 961, Springer-Verlag, Berlin and New York, pp. 314–491.

[VG] L. van Gastel, *Degenerations of conormal varieites*, preprint 439, Dep. Math. Utrecht (1986).

Department of Mathematics, 312 TMCB, Brigham Young University, Provo, Utah 84602